# Lecture Notes in Mathematics



Edited by A. Dold and B. Eckmann

909

# Numerical Analysis

Proceedings of the Third IIMAS Workshop Held at Cocoyoc, Mexico, January 1981

Edited by J.P. Hennart

Springer-Verlag
Berlin Heidelberg New York 1982

**Editor**

J.P. Hennart
IIMAS – UNAM, Apdo. Postal 20-726
01000 México, D.F., Mexico.

AMS Subject Classifications (1980): 65F-XX, 65K05, 65L, 65M, 65N

ISBN 3-540-11193-X Springer-Verlag Berlin Heidelberg New York
ISBN 0-387-11193-X Springer-Verlag New York Heidelberg Berlin



Printing and binding: Beltz Offsetdruck, Hemsbach/Bergstr.
2141/3140-543210

Math

Sep.

# FOREWORD

During the five days 19th-23rd January 1981 in Cocoyoc, Morelos, Mexico, the Institute for Research in Applied Mathematics and Systems (IIMAS) of the National University of Mexico (UNAM) held its Third Workshop on Numerical Analysis. As in the first two versions in 1978 and 1979, the program of this research workshop concentrated on the numerical aspects of three main areas, namely optimization, linear algebra and differential equations, both ordinary and partial. R.H. Bartels, W.C. Davidon, J.E. Dennis Jr., J. Douglas Jr., R. England, C.W. Gear, J.P. Hennart, A.V. Levy, C. Moler, J.J. More, R.W.H. Sargent, R. Tapia and E.L. Wachspress were invited to present lectures. In total, 34 papers were delivered, of which approximately two thirds are given in these Proceedings, reflecting partly the informal aspect of what was a genuine workshop where not all the speakers felt compelled to write down what they presented.

This workshop was supported in part by a generous grant from the Mexican National Council for Science and Technology (CONACyT) and the U.S. National Science Foundation, and was part of the Joint Scientific and Technical Cooperation Program existing between these countries. In relation to this important funding aspect, it is a pleasure to express my thanks to R. Tapia, chairman of the Mathematical Sciences Department at Rice, for his continual advice and help prior to the workshop.

My thanks also go to IIMAS for its generous support and to my colleagues of the Numerical Analysis Department for their friendly cooperation. Finally, the help of the Centre de Mathématiques Appliquées at the Ecole Polytechnique in Palaiseau in the last stages of the editing procedure is gratefully acknowledged.

Palaiseau, October 1981

J.P. HENNART

CONTENTS

P.T. BOGGS and J.W. TOLLE : Merit functions for nonlinear programming problems 1

J.C.P. BUS : Global convergence of Newton-Like methods 11

A.V. LEVY , A. MONTALVO, S. GOMEZ and A. CALDERON : Topics in global optimization 18

S. GOMEZ and A.V. LEVY : The tunnelling method for solving the constrained global optimization problem with several non-connected feasible regions 34

R.H. BARTELS and A.R. CONN : An approach to nonlinear $\ell_1$ data fitting 48

J.L. FARAH : Towards an unified approach to data smoothing 59

A.K. CLINE, A.R. CONN and C.F. VAN LOAN : Generalizing the LINPACK condition estimator 73

C. MOLER : Demonstration of a matrix laboratory 84

M.L. OVERTON : A fast algorithm for the Euclidean distance location problem 99

E.L. WACHSPRESS : Discrete pressure equations in incompressible flow problems 106

S. KAUFMANN and A. MONTALVO : Standing waves in diffusive reacting systems 116

B. CHEN and A. NOYOLA : A study of the stability of the interface between two inmiscible viscous fluids 125

J. NOCEDAL : Solving large nonlinear systems of equations arising in mechanics 132

G. PAGALLO and V. PEREYRA : Smooth monotone spline interpolation 142

R. ENGLAND : Some hybrid implicit stiffly stable methods for ordinary differential equations 147

W.H. ENRIGHT : Developping effective multistep methods for the numerical solution of systems of second order initial value problems 159

P. NELSON, S. SAGONG and I.T. ELDER : Numerical solution of singular two-point boundary-value problems by invariant imbedding 166

R.D. RUSSELL : Difficulties in evaluating differential equation software 175

J.P. HENNART : Topics in finite element discretization of parabolic evolution problems 185

H. GOURGEON and J.P. HENNART : A class of exponentially fitted piece-wise continuous methods for initial value problems 200

R.W.H. SARGENT : Recursive quadratic programming algorithms and their convergence properties 208

D. GOLDFARB and A. IDNANI : Dual and primal-dual methods for solving strictly convex quadratic programs 226

I.S. DUFF : The design and use of a frontal scheme for solving sparse unsymmetric equations 240

## LIST OF PARTICIPANTS

| | |
|---|---|
| BARTELS, R.H. | Computer Science Department, University of Waterloo, Waterloo, Ontario N2L 3G1, Canada |
| BOGGS, P.T. | U.S. Army Research Office, Research Triangle Park, North Carolina 27709, USA |
| BUS, J.C.P. | Stichting Mathematisch Centrum, Kruislaan 413, 1098 SJ Amsterdam, The Netherlands |
| CALDERON, A. | IIMAS-UNAM, Apdo. Postal 20-726, 01000 México D.F., Mexico |
| CHEN, B. | IIMAS-UNAM, Apdo. Postal 20-726, 01000 México D.F., Mexico |
| DAVIDON, W.C. | Department of Physics, Haverford College, Haverford, Pennsylvania 19041, USA |
| DENNIS, Jr., J.E. | Mathematical Sciences Department, Rice University, Houston, Texas 77001, USA |
| DOUGLAS, Jr., J. | Department of Mathematics, The University of Chicago, 5734 University Avenue, Chicago, Illinois 60637, USA |
| DUFF, I.S. | Computer Science and Systems Division, AERE Harwell, Didcot, Oxon OX11 ORA, England |
| ENGLAND, R. | IIMAS-UNAM, Apdo. Postal 20-726, 01000 México D.F., Mexico |
| ENRIGHT, W.H. | Department of Computer Science, University of Toronto, Toronto M5S 1A7, Canada |
| FARAH, J.L. | IIMAS-UNAM, Apdo. Postal 20-726, 01000 México D.F., Mexico |
| GAY, D.M. | MIT-CCREMS, Cambridge, Massachusetts 02139, USA |
| GEAR, C.W. | Department of Computer Science, University of Illinois, Urbana-Champaign, Illinois 61801, USA |
| GOLDFARB, D. | The City College of New York, CUNY, New York, N.Y. 10031, USA |
| GOMEZ, S. | IIMAS-UNAM, Apdo. Postal 20-726, 01000 México D.F., Mexico |
| GOURGEON, H. | IIMAS-UNAM, Apdo. Postal 20-726, 01000 México D.F., Mexico |
| HENNART, J.P. | IIMAS-UNAM, Apdo. Postal 20-726, 01000 México D.F., Mexico |
| HERRERA, I. | IIMAS-UNAM, Apdo. Postal 20-726, 01000 México D.F., Mexico |
| KAUFMANN, S. | IIMAS-UNAM, Apdo. Postal 20-726, 01000 México D.F., Mexico |
| LENTINI, M. | Departamento de Matemáticas, Universidad Simon Bolivar, Sartanejas, Estado de Miranda, Venezuela |
| LEVY, A.V. | IIMAS-UNAM, Apdo. Postal 20-726, 01000 México D.F., Mexico |
| MOLER, C. | Department of Computer Science, University of New Mexico, Albuquerque, New Mexico 87131, USA |
| MONTALVO, A. | IIMAS-UNAM, Apdo. Postal 20-726, 01000 México D.F., Mexico |

| | |
|---|---|
| MORE, J.J. | Division of Applied Mathematics, Argonne National Laboratory, 9700 South Cass Avenue, Argonne, Illinois 60439, USA |
| NELSON, P. | Department of Mathematics, Texas Tech. University, Lubbock, Texas 79409, USA |
| NOCEDAL, J. | IIMAS-UNAM, Apdo. Postal 20-726, 01000 México D.F., Mexico |
| OVERTON, M. | Courant Institute of Mathematical Sciences, 251 Mercer Street, New York, N.Y. 10012, USA |
| PEREYRA, V. | Escuela de Computación, Facultad de Ciencas, Universidad Central de Venezuela, Caracas, Venezuela |
| RUSSELL, R.D. | Department of Mathematics, Simon Fraser University, Burnaby, British Columbia V5A 1S6, Canada |
| SARGENT, R.W.H. | Chemical Engineering Department, Imperial College, London SW7 2BY, England |
| STEIHAUG, T. | Mathematical Sciences Department, Rice University, Houston, Texas 77001, USA |
| TAPIA, R. | Mathematical Sciences Department, Rice University, Houston, Texas 77001, USA |
| VAN LOAN, C.F. | Department of Computer Science, Cornell University, Ithaca, New York 14853, USA |
| WACHSPRESS, E.L. | General Electric Company, Knolls Atomic Power Laboratory, Schenectady, New York 12301, USA |
| WALKER, H.F. | Department of Mathematics, University of Houston, Houston, Texas 77004, USA |

# MERIT FUNCTIONS FOR NONLINEAR PROGRAMMING PROBLEMS[1]

by

Paul T. Boggs[2] and Jon W. Tolle[3]

1. This work was supported in part by ARO Grant DAAG29-79-G0014.

2. U.S. Army Research Office, Research Triangle Park, North Carolina 27709 and Curriculum in Operations Research and Systems Analysis, University of North Carolina, Chapel Hill, North Carolina 27514.

3. Department of Mathematics and Curriculum in Operations Research and Systems Analysis, University of North Carolina, Chapel Hill, North Carolina 27514.

## 1. Introduction

Algorithms for nonlinearly constrained optimization problems have recently been the subject of considerable research activity, stemming in large measure from the application of the quasi-Newton techniques which have been so successful in the unconstrained case. This approach has been popularized by such authors as Han [7], [8], Powell [9], [10] and Tapia [11], [12] among others. While much has been accomplished in the last few years, there are still many unanswered questions relative to both the theory and to the numerical implementation of these ideas. In this paper, we suggest some techniques for the implementation of a quasi-Newton method which relate to assessing the steps which are generated and to the potential for creating a global strategy. By a global strategy, we mean a procedure for forcing convergence from a remote starting value.

We take the problem to be

$$\text{(NLP)} \qquad \begin{aligned} &\min f(x) \\ &\text{subject to:} \quad g(x) = 0 \end{aligned}$$

where $f$ and $g$ are smooth functions, $f : \mathbb{R}^N \to \mathbb{R}^1$ and $g : \mathbb{R}^n \to \mathbb{R}^m$, $m < n$. The quasi-Newton algorithm is easily derived by recalling that the solution to (NLP) can be characterized in terms of the classical Lagrangian function denoted by

$$\ell(x,\lambda) = f(x) + g(x)^T\lambda$$

where $\lambda \in \mathbb{R}^m$ is the vector of Lagrange multipliers. It is well known that if a point $x^*$ is a solution to (NLP) with corresponding multipliers $\lambda^*$, then $(x^*,\lambda^*)$ also satisfies

$$\nabla\ell(x^*,\lambda^*) = \begin{bmatrix} \nabla f(x^*) + \nabla g(x^*)\lambda^* \\ g(x^*) \end{bmatrix} = 0 \quad .$$

Thus, following Tapia [12], we suppose that we have a current approximation $(x^c,\lambda^c)$ to $(x^*,\lambda^*)$ and we apply a structured quasi-Newton method to the system $\nabla\ell = 0$. This yields the system

$$(1.1) \qquad \begin{bmatrix} B & \nabla g(x^c) \\ \nabla g(x^c)^T & 0 \end{bmatrix} \begin{bmatrix} s_x \\ s_\lambda \end{bmatrix} = - \begin{bmatrix} \ell_x(x^c,\lambda^c) \\ g(x^c) \end{bmatrix}$$

where $B$ is an approximation to $\ell_{xx}(x^c,\lambda^c)$. (Here and henceforth the subscript $x$ in $\ell_x$ implies partial derivative. Thus $\ell_{xx}$ is the Hessian with respect to $x$ of the Lagrangian.) The solution of (1.1) provides the steps $s_x$ in $x$ and $s_\lambda$ in $\lambda$ and the next approximation is given by

$$x^+ = x^c + s_x$$

(1.2)

$$\lambda^+ = \lambda^c + s_\lambda \quad .$$

If the point $(x^+,\lambda^+)$ is "acceptable" then $(x^+,\lambda^+)$ replaces $(x^c,\lambda^c)$ and the process is repeated. (In the sequel, we will refer to the sequence of points so generated as $\{x^k\}$ and $\{\lambda^k\}$ and the sequence of steps by $\{s_x^k\}$ and $\{s_\lambda^k\}$.) Of course, in any implementation, one must have a decision procedure for accepting a step and a means for modifying, or shortening, $s_x$ and $s_\lambda$ when the step is not accepted. One purpose of this paper is to suggest a procedure for determining acceptability. The procedure is such that full steps will be acceptable near the solution.

We note at this point that the basic algorithm implied by (1.2) yields equivalent steps to the so-called "recursive quadratic programming" algorithm which replaces $f$ by a quadratic function and $g$ by a linear approximation at each step. (See Tapia [12].)

Returning to our development, our decision procedure is based on the existence of a "merit function," $\phi(x)$, which has the properties:

1. $s_x$ is a descent direction for $\phi(x)$, i.e., there is an $\alpha^* > 0$ such that $\phi(x^c + \alpha s_x) < \phi(x^c)$, $0 < \alpha < \alpha^*$; and
2. $\phi(x^c + s_x) < \phi(x^c)$ when $||x^c - x^*||$ is sufficiently small.

Property 2 allows for Q-superlinear convergence, a point which we discuss in more detail in section 2. It turns out that we will not be totally successful in satisfying property 1, but we are able to suggest a procedure for overcoming the problem.

In section 2 we develop a candidate merit function and derive some of its properties. In section 3 we discuss and justify our basic implementation and present some numerical results. Before proceeding, however, we formalize our notation.

For the problem (NLP) considered here we assume $f$ and $g$ are at least three times continuously differentiable and that the gradient $\nabla g(x)$ has full rank for all $x$. In addition we assume that (NLP) has a (local) solution $x^*$ at which the second order sufficiency conditions hold. That is, there exists a unique vector $\lambda^* \in \mathbb{R}^m$ such that

(i) $\ell_x(x^*,\lambda^*) = 0$

(ii) $\nabla g(x^*)^T y = 0$, $y \neq 0$ imply $y^T \ell_{xx}(x^*,\lambda^*)y > 0$ .

For functions $h : \mathbb{R}^n \to \mathbb{R}^q$, we denote the Jacobian and Hessian matrices by $\nabla h(x)$ and $\nabla^2 h(x)$, respectively. Here, for notational convenience, $\nabla h(x)$ is always written as an $n \times q$ matrix. For functions of $x$ and $\lambda$, we denote derivatives with respect to $x$ or $\lambda$ by subscripts; hence, $\ell_x(x,\lambda) = \nabla f(x) + \nabla g(x)\lambda$, $\ell_{x\lambda}(x,\lambda) = \nabla g(x)$, etc.

Vectors are always column vectors unless transposed, the tranposition operation for vectors and matrices being indicated by a superscript $T$.

The sequence of $B$'s which approximate $\ell_{xx}(x^*,\lambda^*)$ will be denoted by $\{B_k\}$

corresponding to the sequence $\{x^k\}$. We assume that $\{B_k\}$ satisfies the following:

(i) Each $B_k$ is symmetric and nonsingular and the sequence is uniformly bounded; and

(ii) The sequence is uniformly positive definite on the null space of $\nabla g(x^*)^T$.

Finally, we often omit the arguments of the various functions if no confusion can arise.

## 2. A Family of Merit Functions

We derive our merit functions using the theory of augmented Lagrangians as developed in Boggs and Tolle [1]. In particular, we define an augmented Lagrangian

$$M(x,\lambda;d) = \ell(x,\lambda) - (d/2)\, \ell_x(x,\lambda)^T Q(x) \ell_x(x,\lambda)$$

where $d$ is a positive scalar parameter and $Q(x)$ is a positive semidefinite matrix for each $x$. As is the case for the standard Lagrangian, the original problem (NLP) is equivalent to

$$\min_x \max_\lambda M(x,\lambda;d) \quad .$$

However, $M$ has the important additional property that it is a concave quadratic function in $\lambda$ and hence can be explicitly maximized for each value of $x$. By performing this maximization, we obtain the equation

$$\arg\max_\lambda M(x,\lambda;d) = \lambda_d(x) = (\nabla g^T Q \nabla g)^{-1}(g/d - \nabla g^T Q \nabla f)$$

which defines $\lambda_d(x)$. It turns out that certain "natural" choices of $Q$ occur; these are discussed in [1]. For our purposes, we choose

$$Q(x) = \nabla g(\nabla g^T \nabla g)^{-1} \nabla g^T$$

and then define our merit function to be the exact penalty function given by

$$\phi_d(x) = M(x,\lambda_d(x);d) \quad .$$

(We remark that $\phi_d$ was also derived by Fletcher [6], although in an entirely different manner.) The following result concerning $\phi_d$ suggests its potential as a merit function.

<u>Theorem 2.1</u>. (Boggs and Tolle [1]) For $d$ sufficiently small, the function $\phi_d$ satisfies

$$\nabla \phi_d(x^*) = 0$$

$$\nabla^2 \phi_d(x^*) \text{ positive definite.}$$

This result states that $x^*$ is a strong local unconstrained minimum of $\phi_d$. The following partial converse makes $\phi_d$ even more attractive.

Theorem 2.2. If a point $x^*$ is a strong local minimum of $\phi_d$ and $g(x^*) = 0$, then $x^*$ is a local solution to (NLP) and the second order sufficiency conditions hold.

Sketch of Proof: Explicitly form $\nabla\phi_d$ and $\nabla^2\phi_d$. Then, by noting that $\lambda_d(x^*) = \lambda^*$, it follows that

$$\nabla\phi_d(x^*) = \ell_x(x^*,\lambda_d(x^*)) = 0 .$$

The second order sufficiency conditions follow from the fact that

$$v^T\ell_{xx}(x^*,\lambda_d(x^*))v = v^T\nabla^2\phi_d(x^*)v > 0$$

for $v \in \mathbb{R}^n$ satisfying $\nabla g(x^*)^T v = 0$, $v \neq 0$.

Theorems 2.1 and 2.2 imply that an algorithm which reduces $\phi_d$ and simultaneously reduces $||g||$ should be an effective general procedure. We now give a result which shows that the quasi-Newton step, $s_x$, reduces $\phi_d$ locally. By a "local descent direction" we mean a step $s$ such that $\phi_d(x + \alpha s) < \phi_d(x)$ for $0 < \alpha < \alpha^*$ when $||x - x^*||$ is sufficiently small.

Theorem 2.3. Let the matrices $B_k$ be positive definite. Then the quasi-Newton step $s_x$ is a local descent direction for $\phi_d$ sufficiently small.

Sketch of Proof: It can be easily shown that $s_x$ is a descent direction when $g(x) = 0$ and thus by continuity, it is a local descent direction.

Actually Theorem 2.3 could be strengthened to say that $s_x$ is a descent direction sufficiently near feasibility.

Now, in order to ascertain the effect of the use of $\phi_d$ on Q-superlinear convergence, we first need to recall a result of Boggs, Tolle, and Wang [2] which gives a characterization of such a rate.

Theorem 2.4. (Boggs, Tolle, Wang [2]) Define the projection

$$P = I - \nabla g(\nabla g^T \nabla g)^{-1}\nabla g$$

and let $P_k \equiv P(x^k)$. Let $\{B^k\}$ satisfy the conditions of section 1 and let $x^k \to x^*$ linearly. Then $x^k \to x^*$ Q-superlinearly if and only if

$$\frac{||P_k(B_k - \ell_{xx}(x^*,\lambda^*))s_x^k||}{||s_x^k||} \to 0 .$$

We observe that the conditions on $\{B_k\}$ seem very reasonable and, in fact, are satisfied if $\ell_{xx}(x^*,\lambda^*)$ is positive definite and $B_k$ is approximated by a standard quasi-Newton method such as the BFGS. In general, however, it is not known how to generate such a sequence of $B_k$'s. We also note that this result seems to be the natural generalization of the Dennis-More characterization of Q-superlinear convergence in the unconstrained case. (See Dennis and Moré [5].)

We now give our result on the use of $\phi_d$ in the presence of Q-superlinear convergence.

Theorem 2.5. Assume $x^k \to x^*$ Q-superlinearly. Then for $d$ sufficiently small and $k$ sufficiently large

$$\phi_d(x^k + s_x^k) < \phi(x^k) .$$

Sketch of Proof. The difference $\phi_d(x^k + s_x^k) - \phi(x^k)$ must be very carefully calculated with the terms involving $d$ being explicitly accounted for. Then, as a result of Theorem 2.4, certain terms tend to zero as $k \to \infty$ and the rest are negative if $d$ is sufficiently small.

Given these results, it would seem reasonable that $\phi_d$ would at least be a candidate for a merit function in a general purpose program for solving (NLP). In the next section, we suggest an implementation of the quasi-Newton method using a variation of $\phi_d$. Our procedure also includes a suggestion for creating a global strategy in the sense of improving the performance for remote starting values.

## 3. Implementation and Results

Given the results in the previous section, it would seem natural to somehow estimate an appropriate value of $d$ and to then use $\phi_d$ as a merit function. Unfortunately, the theory does not give any computationally reasonable means of doing this. However, the theory together with the actual form of $\phi_d$ allows us to suggest an implementable procedure.

After some simplification, we obtain

$$\begin{aligned}\phi_d(x) &= (1/d)g(x)^T(\nabla g(x)^T\nabla g(x))^{-1}g(x) + \ell(x,\bar{\lambda}(x)) \\ &= (1/d)\phi_1(x) + \phi_2(x)\end{aligned}$$

which defines $\phi_1$ and $\phi_2$. Here, $\bar{\lambda}(x)$ is the least squares approximation to the Lagrange multipliers given by

$$\bar{\lambda}(x) = -(\nabla g(x)^T\nabla g(x))^{-1}\nabla g(x)^T\nabla f(x) .$$

Note that both $\phi_1$ and $\phi_2$ are independent of $d$.

Now, since the results of the previous section hold only when $d$ is sufficently small, we observe that a reduction in $\phi_1$ alone will suffice to allow our step to be accepted. We observe further that $\phi_1$ is a (scaled) measure of the non-feasibility of $x$ and hence that near or at feasibility, it may be difficult or impossible to reduce $\phi_1$. Hence, from such points, $\phi_2$ must be decreased. Based on these comments, we outline our basic algorithm. The details are beyond the scope of this paper and may be found in Boggs and Tolle [3].

1. Generate $(s_x, s_\lambda)$ at $(x^c, \lambda^c)$ using (1.1) and form the trial step $x^+ = x^c + \alpha s_x$, $\lambda^+ = \lambda^c + \alpha s_\lambda$. Here we assume that we have a procedure to bound $||(s_x, s_\lambda)||$ and that $\alpha$ has been chosen accordingly.
2. If $\phi_1(x^+) < \phi_1(x^c)$ then

$$\begin{aligned}&\text{accept } \alpha s_x\\ &x^c := x^+\\ &\lambda^c := \lambda^+\\ &\text{go to } 1\,.\end{aligned}$$

3. If $\phi_1(x^c) < \varepsilon$ then

$$\begin{aligned}&\text{if } \phi_2(x^+) < \phi_2(x^c) \text{ then}\\ &\qquad \text{accept } \alpha s_x\\ &\qquad x^c := x^+\\ &\qquad \lambda^c := \lambda^+\\ &\qquad \text{go to } 1\,.\end{aligned}$$

4. Reject step. (If a step is rejected, we reduce $\alpha$ and try again until it appears that $s_x$ is not a suitable direction.)

Recall that our decision procedure depends on $s_x$ and not at all on $s_\lambda$. Note further that this procedure always accepts the step if $\phi_1$ is reduced and only tries $\phi_2$ when $x^c$ is near feasibility. While it may seem that we have replaced the problem of choosing $d$ by the problem of choosing $\varepsilon$, the latter choice is much easier. Small values of $\varepsilon$ lead to an algorithm which follows the constraints very closely. Larger values allow much less restrictive steps. A value of .1 has performed quite satisfactorily in practice, in fact, much better than smaller values. The use of values such as .01 or .001 seriously degrade the performance especially in cases with highly nonlinear constraints.

The algorithm in this form is closely related to the newly proposed "watchdog" strategy of Chamberlain, Lemarechal, Pedersen and Powell [4] in which they employ two merit functions. These are the Han-Powell choice of

$$\theta(x) = f(x) + \rho \sum_{i=1}^{m} |g_i(x)| \,, \tag{3.1}$$

where $\rho$ is a suitably chosen scalar, and the standard Lagrangian. The decrease of either of them is the criterion for acceptance. The use of $\theta(x)$ is included to force global convergence (see Han [7]) while the use of the Lagrangian allows superlinear convergence. (The authors show that $\theta(x)$ by itself is, in general, incompatible with such a rate.) Our numerical results indicate that our merit function leads to a less conservative algorithm, especially from remote starting values, and thus it appears to warrant further consideration for use in a general purpose program.

We have programmed our algorithm and have tried it on a standard set of test problems found in the literature. Again, the details are beyond the scope of this paper and a complete description may be found in [3]. We present here some representative numerical results and formulate some tentative conclusions.

Table 1 contains a typical sample of results which we obtained. In this table the results using $\phi_d$ are headed by "$\phi_d$" while those using the Han-Powell merit function (3.1) are headed by "H-P". The letter "C" means convergence was obtained, "F" means failure to converge. The number in the #Eval column is the number of evaluations each of $f$, $\nabla f$, $g$ and $\nabla g$.

The results indicate that our method is usually superior to the Han-Powell merit function, i.e., it often allows a larger step to be taken and this extra freedom is warranted. In fact, the test program was constructed in such a way that both merit functions were evaluated at each step and a count was made of the number of times one indicated acceptance when the other indicated rejection. On an enlarged set of test problems, this count was 247 for $\phi_d$ over H-P versus 128 for H-P over $\phi_d$. This ratio of almost 2:1 would have been almost 4:1 except for one of the test problems.

One important feature of this procedure is that it admits a natural global strategy. From Theorem 2.2, we recall that any local minimum of $\phi_d$ which occurs at a feasible point will be a solution to (NLP). Thus, if our line search fails and $||g||$ is large, then a reduction in $\phi_d$ implies a reduction in $\phi_1 = g^T(\nabla g^T \nabla g)^{-1} g$. A direction which reduces $\phi_1$ is not obvious, but a step which reduces $g^T g$ is very easy to generate. In fact,

$$\hat{s} = -\nabla g(\nabla g^T \nabla g)^{-1} g \tag{3.2}$$

is always a descent direction for $g^T g$. Clearly, if $g^T g$ is reduced sufficiently, then $\phi_1$ must eventually be reduced as well. Thus, a natural global strategy for the present algorithm can be to use the quasi-Newton method until a failure is reached and then switch to a procedure which uses (3.2) to reduce $\phi_1$. This idea was tried with the result that on a set of ten problems on which failure occurred when using $\phi_d$, all were eventually solved by this technique. (See [3] for the details.)

| | $\phi_d$ | | H-P | |
|---|---|---|---|---|
| PROBLEM | Converge | # Eval. | Converge | # Eval. |
| 1.1 | C | 9 | F | - |
| 1.2 | C | 17 | C | 17 |
| 1.3 | C | 32 | C | 26 |
| 2.1 | C | 14 | C | 44 |
| 2.2 | C | 22 | C | 36 |
| 3.1 | C | 8 | F | - |
| 4.1 | C | 16 | F | - |
| 5.1 | C | 13 | F | - |
| 6.1 | F | - | F | - |
| 7.1 | C | 23 | F | - |
| 8.1 | C | 16 | F | - |
| 9.1 | F | - | C | 35 |

## REFERENCES


[1] Boggs, P. T. and Tolle, J. W., "Augmented Lagrangians Which are Quadratic in the Multiplier," Journal of Optimization Theory and Applications, Vol. 31, 1980, pp. 17-26.

[2] Boggs, P. T., Tolle, J. W. and Wang, P., "On the Local Convergence of Quasi-Newton Methods for Constrained Optimization," to appear in SIAM Journal on Control and Optimization.

[3] Boggs, P. T. and Tolle, J. W., "An Implementation of a Quasi-Newton Method for Constrained Optimization," Operations Research and Systems Analysis Technical Report No. 81-3, University of North Carolina, Chapel Hill, NC, 1981.

[4] Chamberlain, R. M., Lemarechal, E., Pedersen, H.C. and Powell, M.J.D., "The Watchdog Technique for Forcing Convergence in Algorithms for Constrained Optimization," Tenth International Symposium on Mathematical Programming, August, 1979.

[5] Dennis, J., and Moré, J., "Quasi-Newton Methods, Motivation and Theory," SIAM Review, Vol. 19, 1977, 46-89.

[6] Fletcher, R., "A Class of Methods for Nonlinear Programming, III: Rates of Convergence," Numerical Methods for Nonlinear Optimization, Edited by F. A. Lootsma, Academic Press, New York, New York, 1972.

[7] Han, S. P., "A Globally Convergent Method for Nonlinear Programming," Journal of Optimization Theory and Applicatons, Vol. 22, 1977, pp. 297-309.

[8] Han, S. P., "Dual Variable Metric Algorithms for Constrained Optimization," SIAM Journal on Control and Optimization, Vol. 15, 1977, 546-565.

[9] Powell, M.J.D., "A Fast Algorithm for Nonlinearly Constrained Optimization Calculations," 1977 Dundee Conference on Numerical Analysis, June 1977.

[10] Powell, M.J.D., "The Convergence of Variable Metric Methods for Nonlinearly Constrained Optimization Calculations," Nonlinear Programming 3, O. Mangasarian, R. Meyer, and S. Robinson, eds., Academic Press, New York, 1978, pp. 27-63.

[11] Tapia, R. A., "Diagonalized Multiplier Methods and Quasi-Newton Methods for Constrained Optimization," Journal of Optimization Theory and Applications, Vol. 22, 1977, 135-194.

[12] Tapia, R. A., "Quasi-Newton Methods for Equality Constrained Optimization: Equivalence of Existing Methods and a New Implementation," Nonlinear Programming 3, O. Mangasarian, R. Meyer, S. Robinson, eds., Academic Press, New York, 1978, pp. 125-164.

# GLOBAL CONVERGENCE OF NEWTON-LIKE METHODS

Jacques C.P. Bus
Mathematical Centre
Amsterdam

ABSTRACT

In this paper we consider a general class of Newton-like methods for calculating the solution of n nonlinear equations in n variables, which are continuously differentiable.

Assuming nonsingularity and Lipschitz continuity of the jacobian (the matrix of first partial derivatives of the system) on a certain level set, then we can derive a global convergence theorem for iterative methods in the given class.

## 1. INTRODUCTION

In this paper we consider functions

$$F: D \to \mathbb{R}^n,$$

where $D \subset \mathbb{R}^n$ is some open set and with F continuously differentiable on D. We call $F'(x)$ the *jacobian* of F at x and denote it by $J(x)$. We study the convergence behaviour of a class of iterative methods which calculate an approximation to some solution $x^* \in D$, i.e. $F(x^*) = 0$. We assume that an initial guess $x_0$ to $x^*$ is given. A basic method to approximate $x^*$ is Newton's method, given by the iteration

$$x_{k+1} = x_k - \lambda_k (J(x_k))^{-1} F(x_k), \tag{1.1}$$

where $\lambda_k \in (0,1]$ is the step length factor which has to be determined to satisfy the inequality

$$\|F(x_{k+1})\| \le \|F(x_k)\|. \tag{1.2}$$

It is well known that asymptotic convergence of Newton's method with $\lambda_k = 1$ for all k is quadratic. Moreover, KANTOROVICH & AKILOW [1964] state a semi-local convergence result for this method and global convergence of the general method (1.1) is discussed in DEUFLHARD [1974a/b]. We speak about Newton-like methods if in (1.1) $(J(x_k))^{-1}$ is replaced by some approximation. For such methods local and semi-local convergence is discussed in detail, see e.g. ORTEGA & RHEINBOLDT [1970], DENNIS & MORÉ [1977] and DEUFLHARD & HEINDL [1979]. In BUS [1980] a global convergence theorem is given for the general class of Newton-like methods, which is a generalization of Deuflhard's theorem for Newton's method. In this paper we shall give an improved version. Neglecting the notion of affine invariance (see DEUFLHARD & HEINDL [1979] and BUS [1980]) we are able to give more elegant conditions which do not involve the condition number of the jacobian matrix as is the case in the old version.

This paper is organized as follows. First we give some preliminary notation and lemmas in section 2. The main results are given in section 3. In section 4 we give a short discussion of these results.

PRELIMINARIES

We assume F to be given as in the introduction. Let $M_n$ denote the set of $(n\times n)$-matrices over $\mathbb{R}$. Following the terminology of ORTEGA & RHEINBOLDT [1970] we give the following definitions:

2.1 DEFINITION. A *Newton-like process* for F is a stationary iterative process defined by an iteration function $\Psi : D_\Psi \subset D \times M_n \to \mathbb{R}^n \times M_n$ satisfying

$$(2.1)\qquad \Psi(x,H) = (x-\lambda(x,H)HF(x),\ \Phi(x,H)),$$

where $\lambda(x,H) \in (0,1]$, $\Phi(x,H) \in M_n$, for $(x,H) \in D_\Psi$.

2.2. DEFINITION. A *Newton-like method* assigns a Newton-like process to each function F.

So, given a function F and an initial pair $(x_0,H_0)$, a Newton-like method generates a sequence $\{(x_k,H_k)\}_{k=0}^N$ of approximations to $(x^*,(J(x_k))^{-1})$. Note that N may be finite as the iterative process may break down. Examples of Newton-like methods are: Newton's method (with or without step length control), difference Newton or secant update Newton methods.

In the sequel, $\|\ \|$ means the euclidean norm for vectors and the spectral norm for matrices. The following definitions are useful.

2.3. DEFINITION. The *level function* L of F is given by

$$(2.2)\qquad L(x) = \|F(x)\|^2.$$

2.4. DEFINITION. Let L be the level function of F and $x \in D$. Denote $S = \{y \mid y \in D, L(y) \le L(x)\}$. Then the *level set* $S_F(x)$ of F with respect to x is the path-connected component of S which contains x.

The following lemma gives a sufficient condition for existence of a step length factor such that the level function decreases.

2.5. LEMMA. Given F and $(x,H) \in D \times M_n$. Define

(2.3) $z(t) = x - tHF(x)$

and assume that $z(t) \in D$ *for* $t \in [0,t_1)$ for certain $t_1 \in (0,1]$. Let

(2.4) $\phi(t) = L(z(t))$

and

(2.5) $e(x) = \|J(x)H - I\|$.

Then $e(x) < 1$ implies

(2.6) $\phi'(0) \leq -2(1-e(x))L(x) < 0$.

PROOF. This follows trivially from the Cauchy-Schwarz inequality. □

The upper bound (2.6) is sharp in the sense that we can find a function F and a point $x \in D$ such that for any $\varepsilon \in (0,1)$ there exists a matrix $H = H_\varepsilon$ with $e(x) = \varepsilon$ and (2.6) holds with equality sign. However, the condition (2.5) is too strong in the sense that $\phi'(0) < 0$ for all H such that $J(x)HF(x)$ and $F(x)$ lie in a nondegenerate cone, which condition can be satisfied with $e(x) > 1$ (e.g. $H = k(J(x))^{-1}$ for $k > 2$ implies $e(x) = k-1 > 1$ and $\phi'(0) = -2kL(x) < 0$).

Lemma 2.5 forms the basis for the global convergence theorem of the next section as it provides a condition for existence of a step length factor such that the level function decreases enough in each step to prove convergence of the level function value to zero, hence convergence of the iterative process to a solution.

The following standard conditions on F and $x_0$ are used.

2.6. CONDITION. Let $F:D \to \mathbb{R}^n$, $D \subset \mathbb{R}^n$ open, $x_0 \in D$.

(i) F is continuously differentiable on D, $F(x_0) \neq 0$.

(ii) $J(x)$ is nonsingular for all $x \in S_F(x_0)$.

(iii) $S_F(x_0)$ is compact.

(iv) There exists an $\omega(x_0) > 0$ such that for all $y,z \in S_F(x_0)$:

$$\|J(y)(J(z))^{-1} - I\| \leq \omega(x_0)\|y-z\|.$$

Finally we give some standard notation which is used throughout section 3.

2.7. NOTATION. Let F and x satisfy condition 2.6. and let $H \in M_n$ be nonsingular.

$\beta(x) = \sup\{\|(J(y))^{-1}F(x)\| \,|\, y \in S_F(x)\}$,

$\omega(x) = \inf\{\omega \,|\, \|J(y)(J(z))^{-1}-I\| \leq \omega\|y-z\|, \forall y,z \in S_F(x)\}$,

$\alpha(x) = \omega(x)\beta(x)$,

$\gamma(x) = \sup\{\|(J(y))^{-1}\| \,|\, y \in S_F(x)\}$,

$e(x) = \|J(x)H-I\|$,

$$\nu^{(0)}(x) = \tfrac{1}{2}(1 + \frac{e(x)}{2(1-e(x))} \quad \frac{\|F(x)\|}{\beta(x)}),$$

$$\nu^{(1)}(x) = 2e(x) + 1/(2\nu^{(0)}(x)),$$

$$\nu^{(2)}(x) = (1-e(x))/\nu^{(0)}(x),$$

$$c(x)(t) = 1 + \nu^{(0)}(x)t((\alpha(x)t)^2 + \nu^{(1)}(x)\alpha(x)t - \nu^{(2)}(x)),$$

$$\zeta(x) = \frac{\nu^{(1)}(x)}{2\alpha(x)} \; (-1 + \sqrt{1 + 4\nu^{(2)}(x)(\nu^{(1)}(x))^{-2}})$$

Note that $\zeta(x)$ is the smallest positive root of $c(x)(t) = 0$. If we work with $x_k$, for some index k, then we may write $\beta_k$ instead of $\beta(x_k)$ and similarly for the other quantities, including $F_k = F(x_k)$, $J_k = J(x_k)$ and $S_F(x_k) = S_k$.

## 3. GLOBAL CONVERGENCE

The first theorem states existence of a solution and of a differentiable path in the level set of F with respect to the initial point $x_0$, going from $x_0$ to the solution.

3.1. THEOREM. Let F and $x_0$ satisfy condition 2.6. Then there exists a unique differentiable function $p:[0,1] \to S_0$ satisfying

$$(3.1) \qquad F(p(t)) = (1-t)F_0, \quad t \in [0,1].$$

Moreover, p satisfies

$$(3.2) \qquad \begin{aligned} p'(t) &= -(J(p(t)))^{-1}F_0, \qquad t \in [0.1] \\ p(0) &= x_0. \end{aligned}$$

Furthermore, $x^* = p(1)$ is a unique solution of $F(x) = 0$ in $S_0$. The path $\{y | y = p(t),\ t \in [0,1]\} \subset S_0$ is called the *Newton path*.

PROOF. See BUS [1980, thm 2.8]. □

The following theorem states stepwise decrease of Newton-like methods for functions satisfying the standard conditions.

3.2. THEOREM. Let F and $x_0$ satisfy condition 2.6. Let $H \in M_n$ and $e_0 < 1$. Define z(t) by (2.3). Then, for all t satisfying

$$0 \le t \le \min(1, \zeta_0),$$

we have

(i) $z(t) \in S_0$,

(ii) $L(z(t)) \le (c_0(t))^2 L(x_0) \le L(x_0)$.

REMARK: $(c_0(t))^2 = \psi(t)$ satisfies $\psi(0) = 1$, $\psi'(0) = -2(1-e_0)$. Hence, $\psi(t)$ is an upper bounding approximation to $L(z(t))/L(x_0)$ which fits in $t = 0$ and has the derivative suggested by lemma 2.5.

PROOF. The proof follows the lines of the proof of theorem 5.4 in BUS [1980]. We only sketch it roughly to indicate the differences. Using $p: [0,1] \to S_0$ from theorem 3.1 we define for $t,s \in [0,1]$:

$$w(t,s) = p(t) + s(z(t) - p(t)),$$

$$\delta(t) = \sup\{s \mid s \in [0,1];\ w(t,s') \in S_0,\ \forall s' \in [0,s]\},$$

$$\bar{S} = \{x \mid x = w(t,s),\ t \in [0,1],\ s \in [0,\delta(t)]\}.$$

Compactness of $S_0$ and continuity of $F$ yields $\bar{S} \subset S_0$. Choose $t \in [0,1]$ fixed and apply the mean value theorem with respect to s. Then

$$F(w(t,s)) = (1-t)F_0 + \Big(\frac{1}{s}\int_0^s J(w(t,s'))ds'\Big)(w(t,s)-p(t))$$
$$= (1-t)F_0 + \frac{1}{s}\Big[1 + \int_0^s (J(w(t,s'))J_0^{-1} - I)ds'\Big](J_0(w(t,s)-p(t))).$$

Hence

$$\|F(w(t,s))\| \le (1-t)\|F_0\|$$
$$+ \frac{1}{s}\Big(1 + \omega_0 \int_0^s \|w(t,s')-x\|ds'\Big)\|J_0(w(t,s)-p(t))\|.$$

We can prove subsequently

$$\|J_0(w(t,s)-p(t))\| \le ts(\tfrac{1}{2}\alpha_0 t+e_0)\|F_0\|$$

and

$$\|w(t,s)-x_0\| \le t\Big(\beta_0+s\,\frac{e_0}{1-e_0}\,\|F_0\|\Big).$$

This yields the result by using the same arguments as in BUS [1980, thm. 5.4]. □

To obtain the global result we need the following lemma which states that the decrease of the level function is with a factor which is bounded away from 1.

**3.3. LEMMA. Let the conditions of theorem 3.2 be satisfied. Suppose $H_0 \in \mathcal{M}_n$ and $\tau$ satisfy**

$$0 < \tau \le \min\Big(1, \frac{2(1-e_0)^2}{\omega_0\|F_0\|(2\gamma_0(1-e_0)+e_0)}\Big).$$

Then $\tau \le \min(1,\zeta_0-\tau)$ and for all $t$ with

$$\tau \le t \le \min(1,\zeta_0-\tau),$$

we have $z_0(t) = x_0-tH_0F_0 \in S_0$ and

$$L(z_0(t)) \le \Big(1 - \frac{\tau^2}{4}(1-e_0)\Big)^2 L(x_0). \qquad (3.3)$$

PROOF. Note that $\gamma_0$ is finite as $S_0$ is compact and $J(x)$ is nonsingular on $S_0$. We have (see also BUS [1980, thm. 5.7]) using $\nu_0^{(2)} \geq \frac{1}{2}$:

$$\zeta_0 \geq \frac{\nu_0^{(1)}}{2\alpha_0}\left(\frac{4\nu_0^{(2)}(\nu_0^{(1)})^{-2}}{2+2(\nu_0^{(1)})^{-1}\sqrt{\nu_0^{(2)}}}\right)$$

$$= \frac{1-e_0}{\nu_0^{(0)}\alpha_0(2e_0+\frac{1}{2}(\nu_0^{(0)})^{-1}+\sqrt{(1-e_0)/\nu_0^{(0)}})}$$

$$\geq \frac{1-e_0}{\nu_0^{(0)}\alpha_0(2e_0+3)}\ .$$

Moreover

$$\nu_0^{(0)}\alpha_0 \leq \frac{1}{2}(\gamma_0 + \frac{e_0}{2(1-e_0)})\omega_0\|F_0\|.$$

So

$$\zeta_0 \geq \frac{4(1-e_0)^2}{(2\gamma_0(1-e_0)+e_0)\omega_0\|F_0\|} \geq 2\tau$$

and $[\tau, \min(1,\zeta_0-\tau)]$ is nonempty. The rest of the proof is the same as in BUS [1980, thm. 5.7]. □

Using this lemma we get the final result:

3.4. THEOREM. Let the standard conditions (2.6) be satisfied for $F$ and $x_0$. Let $H_0 \in M_n$ be given and $\{(x_k,H_k)\}_{k=0}^{\infty}$ be generated by a Newton-like method. Suppose $e_k < e$ for all $k$ and there exists a number $\sigma \in (0,1)$ such that the step length factor satisfies

(3.4) $\qquad \sigma \leq \lambda(x_k,H_k) \leq \min(1,\zeta_k-\sigma)$

for all $k$. Then $(x_k,H_k)$ is well-defined for all $k$, $\{x_k\}_{k=0}^{\infty} \in S_0$ and converges to a unique point $x^* \in S_0$ with $F(x^*) = 0$. Moreover, there exists an integer $K \geq 0$ such that $\lambda(x_k,H_k) = 1$ satisfies (3.4) for all $k \geq K$.

PROOF. We know that $\gamma_0$ is bounded. Define

$$\tau = \min(\sigma, \frac{2(1-e)^2}{\omega_0\|F_0\|(2\gamma_0(1-e)+e)}).$$

Then, as $\omega_0\|F_0\| \geq \omega_k\|F_k\|$, $\gamma_0 \geq \gamma_k$ for all $k$, we can apply lemma 3.3 in every iteration step with this $\tau$. The rest of the proof is given in BUS [1980, thm. 5.8].

## 4. DISCUSSION

We presented a global convergence result for a general class of Newton-like methods. This result is applicable in all algorithms for which can be proven in advance that the error in the inverse jacobian approximation is small enough ($\| J(x)H - I\| < 1$). If this error can be predicted a priori for a certain approximation method then this prediction can be used to control the error and achieve global convergence. Examples of such algorithms are given in BUS [1980].

## REFERENCES

BUS, J.C.P. [1980], *Numerical solution of systems of nonlinear equations*, Mathematical Centre Tracts 122, Amsterdam.

DENNIS jr., J.E. & J.J. MORÉ [1977], *Quasi-Newton methods, motivation and theory*, SIAM Rev. 19, 46-89.

DEUFLHARD, P., [1974a], *A modified Newton method for the solution of ill-conditioned systems of nonlinear equations with application to multiple shooting*, Numer. Math. 22, 289-315.

DEUFLHARD, P., [1974b], *A relaxation strategy for the modified Newton method* in: Conference on optimization and optimal control, BULIRSCH, R., W. OETTLI & J. STOER (eds), Oberwolfach, Springer, Berlin.

DEUFLHARD, P. & G. HEINDL, [1979], *Affine invariant convergence theorems for Newton's method and extensions to related methods*, SIAM J. Numer. Anal. 16, 1-10.

KANTOROVICH, L.W. & G.P. AKILOW, [1964], *Functional analysis in normed spaces* (german), publ. by P.H. Müller, transl. from Russ. by H. Langer and R. Kühne, Berlin, Akademie-Verlag, Math. Lehrbücher und Monographien: 2.17.

ORTEGA, J.M. & W.C. RHEINBOLDT, [1970], *Iterative solution of nonlinear equations in several variables*, Academic Press, New York & London.

# TOPICS IN GLOBAL OPTIMIZATION

A. V. Levy, A. Montalvo,
S. Gomez and A. Calderon.

IIMAS - UNAM.
A.P. 20-726, Mexico 20, D.F.
Mexico City, Mexico.

ABSTRACT. A summary of the research done in global optimization at the Numerical Analysis Departament of IIMAS-UNAM is given. The concept of the Tunnelling Function and the key ideas of the Tunnelling Algorithm as applied to Unconstrained Global Optimization, Stabilization of Newton's Method and Constrained Global Optimization are presented. Numerical results for several examples are given, they have from one to ten variables and from three to several thousands of local minima, clearly illustrating the robustness of the Tunnelling Algorithm.

## 1. UNCONSTRAINED GLOBAL OPTIMIZATION. ( Refs. 1,2 ).

### 1.1 Statement of the problem.

In this section we consider the problem of finding the global minimum of f(x), where f is a scalar function with continuous first and second derivatives, x is an n-vector, with $A \leq x \leq B$, where A and B are prescribed n-vectors.

In this work we assume that the problem does have a solution and if we say that $f(x^*)$ is the global minimum, this means that a) $f(x^*)$ is a local minimum, satisfying the optimality conditions $f_x(x)=0$ , $f_{xx}(x)$ p.d., and b) at $x^*$ the function has its lowest possible value, satisfying the condition $f(x^*) < f(x)$ , for $A \leq x \leq B$ .

The particular cases, when there are several local minima at the same function level, or the lowest value of the function occurs at the boundary of the hypercube $A \leq x \leq B$, were considered in the original paper, (Ref.2) but for brevity they are omitted in this section. The type of functions f(x) being considered, are illustrated in Figs. 1 and 2.

### 1.2 Description of the Tunnelling Algorithm.

Design Goal. Since the type of problems considered in global optimization have a large number of local minima, the design goal of this algorithm is to achieve a Generalized Descent Property, that is, find sequentially local minima of f(x) at $x_i^*$ , i=1,2,...,G, such that

$$f(x_i^*) \geqslant f(x_{i+1}^*) \quad , \quad A \leqslant x_i^* \leqslant B \quad , \quad i=1,2,\dots,G-1 \tag{1}$$

thus avoiding irrelevant local minima, approaching the global minimum in an orderly fashion.

Structure of the Tunnelling Algorithm. The Tunnelling Algorithm is composed of a sequence of cycles, each cycle consists of two phases, a minimization phase and a tunnelling phase.

The minimization phase is designed to decrease the value of the function. For each given nominal point $x_i^\circ$, i=1,2,.. ,G, find a local minimum say at $x_i^*$ , that is solve the problem

$$\min_x f(x) = f(x_i^*) \quad , \quad i=1,2,\dots,G \tag{2}$$

Any minimization algorithm, with a local descent property on f(x) can be used in this phase.

The tunnelling phase is designed to obtain a good starting point for the next minimization phase. Starting from the nominal point $x_i^*$ , i=1,2,..G-1 we find a zero of the "TunnellingFunction", $T(x,\Gamma)$, that is

$$T(x_{i+1}^\circ,\Gamma) \leqslant 0 \quad , \quad i=1,2,\dots G-1 \tag{3}$$

using any zero finding algorithm.

## 1.3 Derivation of the Tunnelling Function.

We define the Tunnelling Function as follows

$$T(x,\Gamma) = \frac{f(x) - f(x^*)}{\{(x-x^*)^T(x-x^*)\}^\lambda} \tag{4}$$

where $\Gamma$ denotes the set of parameters $\Gamma=(x^*, f(x^*),\lambda)$. Why we arrive at this definition can be seen in the following derivation:

a)-Consider Fig. 3 .a. Let $x^*$ be the relative minimum found in the last minimization phase. We would like to find a point $x^\circ$, $x^\circ \neq x^*$, $A \leqslant x^\circ \leqslant B$ , where

$$f(x^\circ) \leqslant f(x^*) = f^* \tag{5}$$

This point $x^\circ$ would be a very good nominal point for starting the next

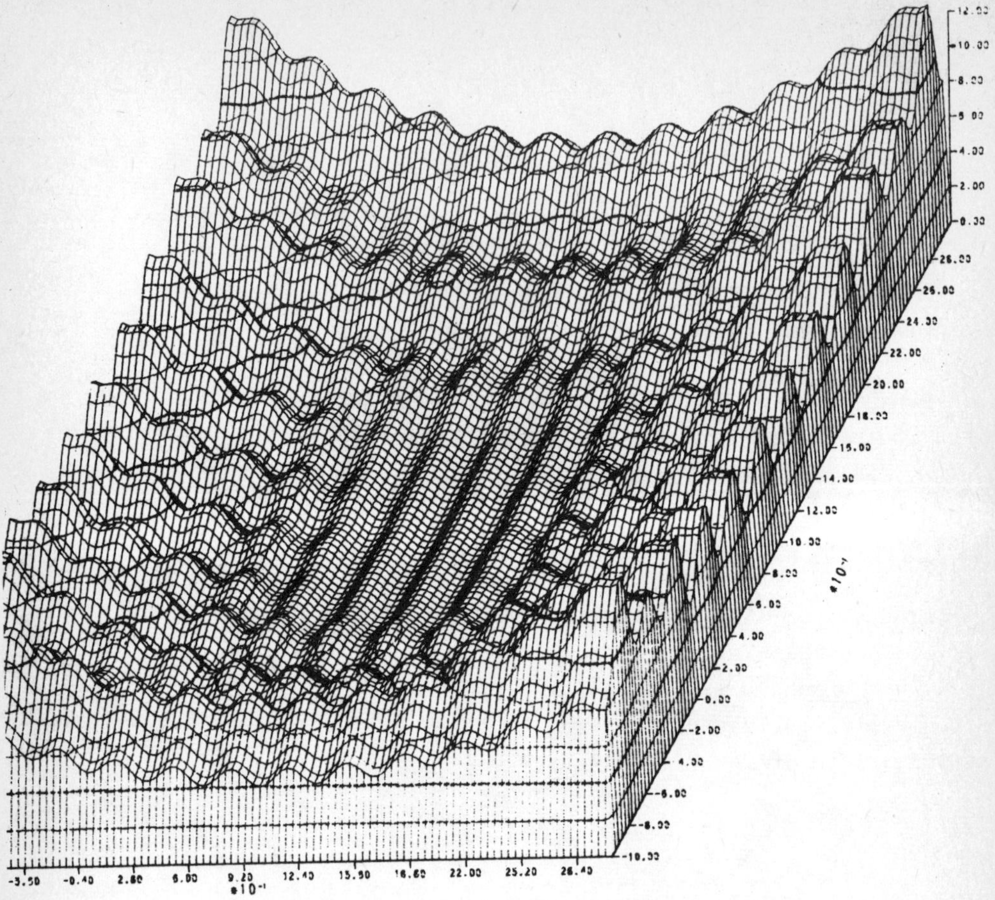

Figure 1. Typical function f(x) being minimized, with 900 local minima. This is a partial illustration of Example 13.

minimization phase, because, obviously, the next minimization phase will produce a local minimum with a lower function value than $f(x^*)$.
So the first thing to do is to store the value of $f(x^*)$ and shift the function, obtaining the expression

$$T(x,f^*) = f(x) - f(x^*) \tag{6}$$

b)-Consider now Fig. 3.b. We have now $T(x^*,f^*) = 0$ , $T(x°,f^*) = 0$ , without having perturbed the position of x°. This being the case, we can find x°, if we look for a zero of $T(x,f^*) = 0$. However to avoid being attracted by the zero at $x^*$, we store the value of $x^*$ and cancell this zero by introducing a pole at $x^*$; thus we obtain the expression

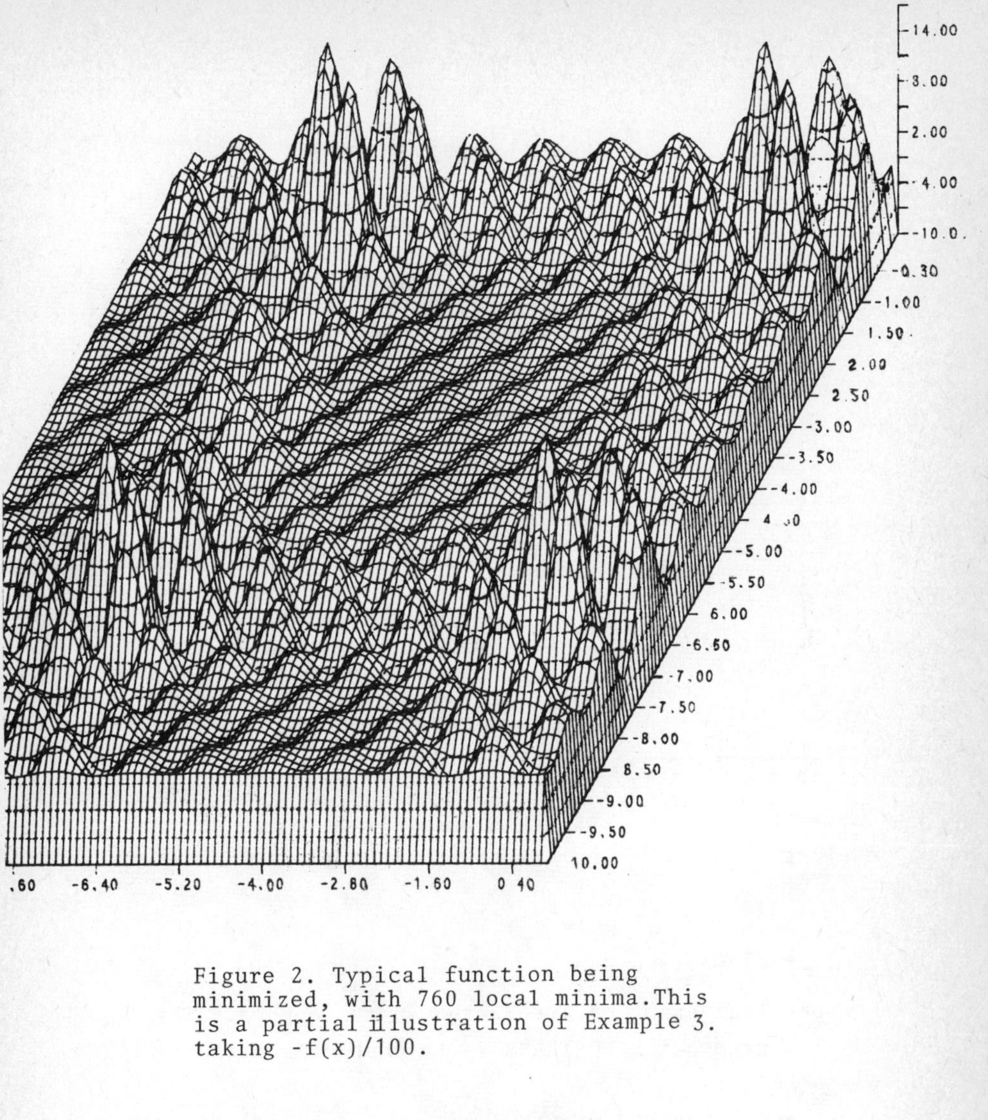


Figure 2. Typical function being minimized, with 760 local minima. This is a partial illustration of Example 3. taking -f(x)/100.

$$T(x,x^*,f^*,\lambda) = \frac{f(x) - f(x^*)}{\{(x-x^*)^T(x-x^*)\}^\lambda} \tag{7}$$

where $\lambda$ denotes the strength of the pole introduced at $x^*$. If we let $\Gamma$ denote the parameters $x^*, f^*, \lambda$ we obtain the definition of the Tunnelling Function given in Eq.(4). Fig. 3 .c illustrates the final effect of mapping $f(x)$ into $T(x,\Gamma)$, once the the zero at $x^*$ has been cancelled.

1.4 <u>Computation of the pole strength</u>.

The correct value of $\lambda$ is computed automatically, starting with $\lambda = 1$ and the value of $T(x,\Gamma)$ at $x=x^*+\varepsilon$ ,where $\varepsilon<<1$ is used just to avoid a division by zero.If $T(x,\Gamma)>0$, $\lambda$ has the correct value.If $T(x,\Gamma)=0$, $\lambda$ is incremented,say by 0.1 until $T(x,\Gamma)>0$.

### 1.5 Generalized Descent Property.

Due to the definition of the minimization and tunneling phases, and their sequential use as shown in Fig.4.(a) the following relationship holds

$$f(x_i^\circ)\geqslant f(x_i^*)\geqslant f(x_{i+1}^\circ)\geqslant f(x_{i+1}^*) \quad ,i=1,2..G-1$$

and dropping the intermediate points $x_i^\circ$ ,the following Generalized Descent Property is obtained

$$f(x_i^*)\geqslant f(x_{i+1}^*) \quad , \; i=1,2,\ldots,G-1 \qquad (8)$$

Having satisfied our design goal for the Tunnelling Algorithm let's look at the numerical experiments.

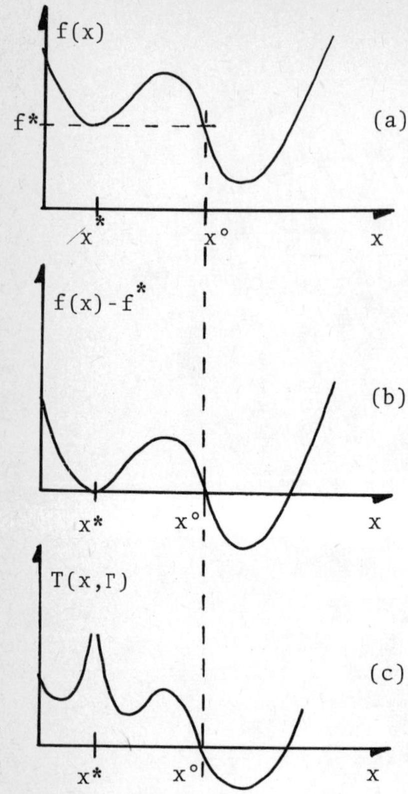


Fig. 3 . Derivation of the Tunneling Function.

### 1.6 Numerical Experiments.

Some of the numerical examples are listed in the appendix.The Tunnelling Algorithm was programmed in FORTRAN IV, in double precision.A CDC-6400 computer was used.

The Ordinary Gradient Method was used in the minimization phase.The stepsize was computed by a bisection process, starting from $\alpha=1$,enforcing a local descent property on $f(\alpha)< f(0)$ .Nonconvergence was assumed if more than 20 bisections were required.The minimization phase was terminated when $f_x^T(x)f_x(x)< 10^{-9}$.

In the tunnelling phase the Stabilized Newton's Method was employed,( see section 2. and Refs. 3, 4).The stopping condition for the tunnelling phase was taken as $T(x,\Gamma) \leqslant 0$.

If in 100 iterations the above stopping condition was not satisfied,it was then assumed that probably the global minimum had been found, and the algorithim was terminated. Obviously, this is only a necessary

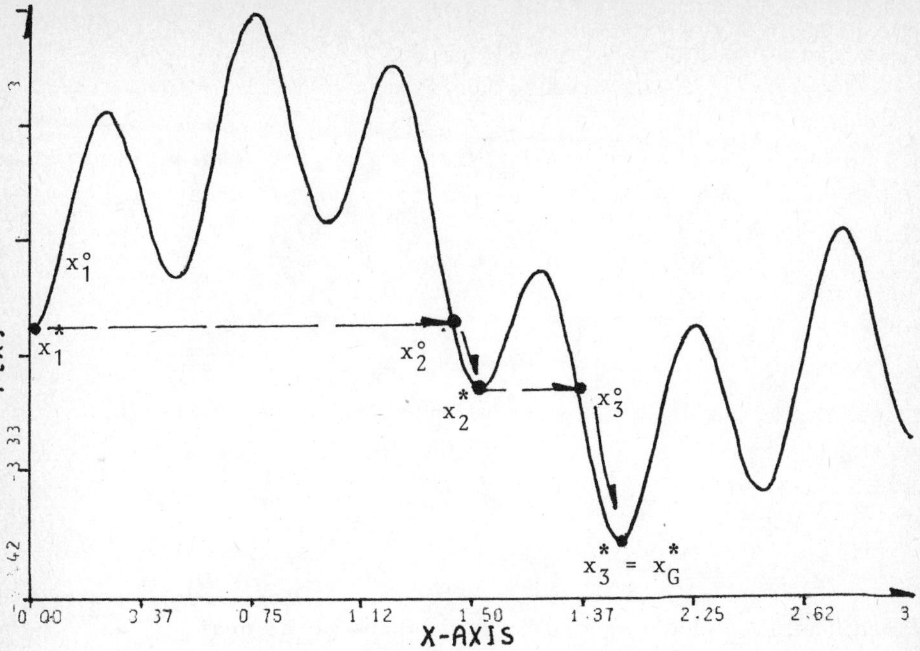


Fig.4a.Geometric interpretation of the Tunnelling Algorithm.The sequential use of Minimization phases and Tunnelling phases,makes the algorithm to ignore local minima whose function value is higher than the lowest $f^*$.

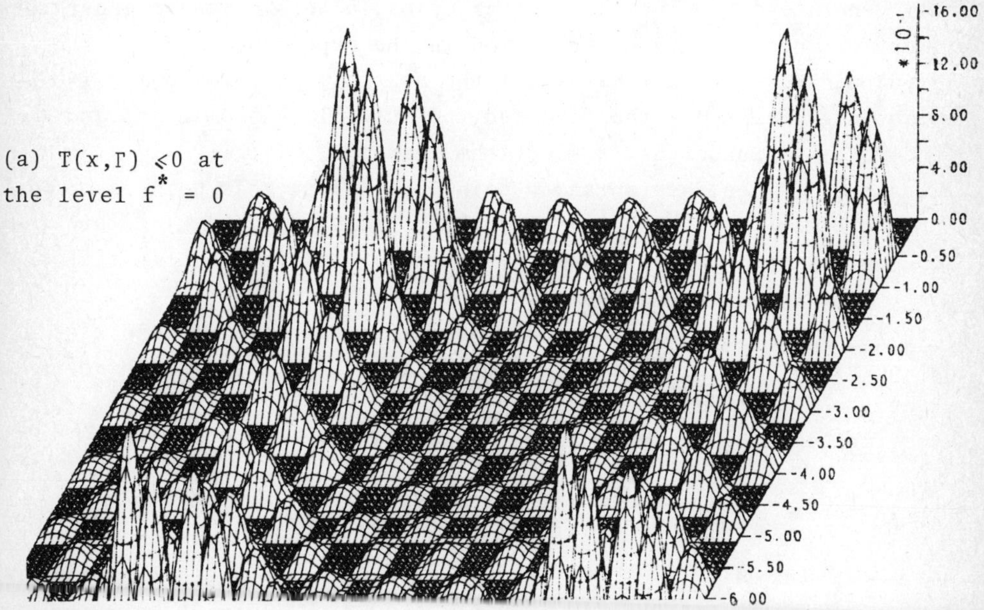


Fig.4b.Typical effect of storing $f^*$ and shifting the original f(x) to obtain $f(x)-f^*$.Irrelevant local minima are ignored,regardless of their position.This figure is a partial illustration of Example 3, taking $-f(x)/100$ and $f^* = 0$

condition for global optimality. To obtain a necessary and sufficient condition for global optimality, instead of a 100 iterations, a very large computing time should be allowed to verify that $T(x,\Gamma) > 0$ for every x in the hypercube.

Each test problem was solved from $N_R$ starting points, chosen at random. If each computer run took $t_i$ CPU seconds to converge, an average computing time can be defined as

$$t_{av} = \frac{\sum_{i=1}^{N_R} t_i}{N_R} \tag{9}$$

If $N_S$ denotes the number of successful runs, the probability of success is given by

$$p = \frac{N_S}{N_R} \tag{10}$$

Since the tunnelling phase starts from a random point near $x^*$, and finds one of the multiple points that can satisfy the stopping condition $T(x^\circ,\Gamma) \leq 0$, one can consider that the tunnelling phase behaves like a "random generator" of points $x^\circ$. To see how well this "random generator" works, we shall generate the points $x^\circ$ by other means, such as a random generator with a uniform probability distribution in the hypercube, and use the generated points as the nominals $x^\circ$ of the next minimization phase. Two comparison algorithms are thus created, the Multiple Random Start and the Modified Multiple Random Start algorithms, as follows ;

MRS Algorithm : a) generate a random point $x^\circ$. b) use this point as nominal for a minimization phase and find the corresponding $x^*$. c) Return to step a), unless the CPU computing time exceeds a given time limit, for example $t_{av}$ of the tunnelling algorithm.

MMRS Algorithm : a) generate a random point $x^\circ$. b) if $f(x^\circ) < f^*$ use $x^\circ$ as a nominal point for a minimization phase and find $x^*$, otherwise do not minimize and go back to step a). c) Stop when the CPU computing time excceds a given time limit, for example the $t_{av}$ of the tunnelling algorithm.

## 1.7 Conclusions

a). The tunnelling algorithm really "tunnels" below irrelevant local minima; from Fig. 5 we see that the number of local minima can go up by a factor of a million, from $10^3$ to $10^8$, while the computing effort goes up only by a factor of ten.

b). The computer runs show that the global minimum is approached in an

| Ex. No. | No. Var. | No. Local Minima | TUNNELLING ALGORITHM | | | | MRS | | MMRS | |
|---|---|---|---|---|---|---|---|---|---|---|
| | | | $N_f$ | $N_g$ | $t_{av}$ | p | $t_{av}$ | p | $t_{av}$ | p |
| 1 | 1 | 3 | 758 | 129 | 0.55 | 1.0 | 0.09 | 1.0 | 0.60 | 0.5 |
| 2 | 1 | 19 | 1502 | 213 | 4.03 | 1.0 | 4.10 | 0.66 | 4.04 | 0.33 |
| 3 | 2 | 760 | 12160 | 1731 | 87.04 | 0.94 | 88.08 | 0.5 | 87.06 | 0.05 |
| 4 | 2 | 760 | 2912 | 390 | 8.47 | 1.0 | 5.19 | - * | 4.31 | - * |
| 5 | 2 | 760 | 2180 | 274 | 5.98 | 1.0 | 2.09 | - * | 6.01 | 0.0 |
| 6 | 2 | 6 | 1496 | 148 | 1.98 | 1.0 | 0.03 | 1.0 | 2.03 | 0.5 |
| 7 | 2 | 25 | 2443 | 416 | 3.28 | 1.0 | 64.0 | - * | 1.06 | 1.0 |
| 8 | 3 | 125 | 7325 | 1328 | 12.91 | 1.0 | 3.51 | 1.0 | 4.02 | 1.0 |
| 9 | 4 | 625 | 4881 | 1317 | 20.41 | 1.0 | 3.39 | - * | 4.33 | 1.0 |
| 10 | 5 | $10^5$ | 7540 | 1122 | 11.88 | 1.0 | 8.10 | - * | 11.92 | 0.0 |
| 11 | 8 | $10^8$ | 19366 | 2370 | 45.47 | 1.0 | 38.09 | 1.0 | 45.53 | 0.0 |
| 12 | 10 | $10^{10}$ | 23982 | 3272 | 68.22 | 1.0 | 192.0 | - * | 68.26 | 0.0 |
| 13 | 2 | 900 | 2653 | 322 | 4.36 | 0.5 | 6.30 | 0.0 | 1.79 | 1.0 |
| 14 | 3 | 2700 | 6955 | 754 | 12.37 | 0.75 | 13.29 | 0.0 | 2.97 | 1.0 |
| 15 | 4 | 71000 | 3861 | 588 | 8.35 | 1.0 | 9.85 | 0.0 | 8.37 | 0.0 |
| 16 | 5 | $10^5$ | 10715 | 1507 | 28.33 | 1.0 | 51.70 | 0.0 | 28.36 | 0.0 |
| 17 | 6 | $10^7$ | 12786 | 1777 | 33.17 | 1.0 | 41.06 | 0.0 | 33.23 | 0.0 |
| 18 | 7 | $10^8$ | 16063 | 2792 | 71.98 | 0.75 | 92.61 | 0.0 | 72.02 | 0.0 |

Fig. 5 Numerical Results. $N_f$ = No. function eval. $N_g$ = No. of gradient eval. $t_{av}$ = average computing time (sec.). p = probability of success. * = Nonconvergence

orderly fashion, verifying in practice the generalized descent property of the tunnelling algorithm.

c). The tunnelling phase is a "very well educated random point generator"; for the same computing effort it provides points $x_i^o$ that are better nominals for the next minimization phase, than those produced by the MRS and MMRS algorithms. This methods have a decreasing probability of success, $p \to 0$ ,as the density of local minima is increased.

d). For problems of small number of variables or with few local minima, a random nominal type algorithm requires less computing effort than the tunnelling algorithm; however, for problems with a large number of local minima, the tunneling algorithm is usually faster and has a higher probability of convergence.

e). Problems with several minima at the same function level are by far the most difficult to solve if all the "global minima" are desired. For instance , Example No. 3 has 18 "global minima" at the same level,(see

Figs. 2,4 and 5) and inspite of having only 760 local minima, the computing effort is similar to problems with $10^8$ local minima. Examples No. 4 and 5 are obtained by from Ex. No.3, by removing 17 of the 18 "global minima", thus becoming much easier to solve.

## 1.8 Convergence Proof of a Modified Tunnelling Algorithm .

In Ref.5, a theoretical proof of the global convergence of a modified tunnelling algorithm towards the global minimum of a one dimensional scalar function is given.

This version of the tunnelling algorithm, uses the same key ideas outlined in the previous sections, as well as the basic structure of sequential phases; a minimization phase and a tunnelling phase. The main modification is in the definition of the tunnelling function, it is defined as

$$T(x,\Gamma) = \frac{\{f(x)-f^*\}^2}{\{(x-x^*)^T(x-x^*)\}^\lambda} \qquad (11)$$

The most important achievements of this modification are: a). It is now possible to establish the theoretical proof of the global convergence of the tunnelling algorithm to the global minimum of the function; thus it is guaranteed that, starting from any nominal point, the global minimum will be found in a finite number of steps. b). The proof of the convergence theorems are constructive proofs and therefore, a practical computer algorithm can be written that implements this global convergence properties. c). To experimentally validate the theoretical convergence theorems, this algorithm was written in FORTRAN IV in single precision and eight numerical examples of a single variable were solved in a B=6700 computer. The numerical results show that the global minimum was always found, regardless of the starting point, thus confirming the global convergence property of this modified tunnelling algorithm.

# 2. STABILIZED NEWTON'S METHOD. ( Refs. 3,4)

## 2.1 Relationship to Global Unconstrained Minimization.

As described in section 1, during the tunnelling phase a zero of the tunnling function $T(x,\Gamma)$ must be found. The function $f(x)$ has usually many local minima. The introduction of the pole at $x^*$ smooths out a few of them, but in general $T(x,\Gamma)$ , will itself have many relative minima.

This means that the tunnelling function has many points $x^s$, where the gradient $T_x(x,\Gamma)$ becomes

$$T_x(x,\Gamma) = 0$$

## 2.2 Statement of the Problem .

In this section we consider the problem of finding a zero of a system of non-linear equations $\Phi(x) = 0$ , where $\Phi$ is a q-vector function with continuos first and second derivatives and x is an n-vector. We assume that there are many singular points $x^s$, where the Jacobian $\Phi_x(x^S)$ is not full rank and $\Phi(x) \neq 0$. We assume a solution exists.

In Ref.2 the particular case of q = 1 was studied with $A \leqslant x \leqslant B$ , x an n-vector. In Ref.4 , the general case $q \leqslant n$ was presented; for brevity we shall only consider in this section the case q=n.

To clarify our notation, let the basic equation in Newton's method be written as

$$\Phi_x^T(x)\ \Delta x = -\ \Phi(x) \tag{12}$$

Solving this system of n equations gives the displacement vector $\Delta x$, and the next position vector is computed as

$$\tilde{x} = x + \beta\Delta x \qquad , \quad 0 < \beta \leqslant 1 \tag{13}$$

where the stepsize $\beta$ is computed so as to enforce a descent property on the norm of the error, $P(x) = \Phi^T(x)\Phi(x)$ , that is, starting from $\beta = 1$, $\beta$ is bisected until the condition

$$P(\tilde{x}) < P(x) \tag{14}$$

is satisfied, accepting then $\tilde{x}$ as the new position vector to start the next solution of Eq. (12).

## 2.3 Description of the Stabilized Newton's Method .

It can be shown that, the damped Newton's Method is attracted to singular points,once it gets close to them, with the stepsize $\beta$ tending to zero as the descent property on P(x) is enforced.Also as given in Refs. 3,4 it can be shown that, for the method to converge, it must be started close to the solution, having a small radius of convergence if singular points are present.

Design Goal. Since we want to solve systems of equations $\Phi(x) = 0$ that have many singular points, our goal is to design an algorithm that stabilizes the Damped Newton's Method, increasing its convergence radius to infinity, making possible for the nominal point to be very far from the solution , even if many singular points are present.

Structure of the Stabilized Newton's Method . The design goal can be achieved as follows; a). detecting if the method is approaching a singular point and b). eliminating the singularity, so that the method is no longer attracted to it.

The stabilized algorithm has two possible phases; in one phase the zero of the original system $\Phi(x)$ is sought, however, when a singularity is detected the method will enter the other phase, were it will seek the zero of an equivalent system $S(x)=0$. For this structure to operate the system $S(x)$ must have the following properties;

$$\text{(a)} \quad \Phi(x^{\circ})=0 \leftrightarrow S(x^{\circ})=0 \tag{15}$$
$$\text{(b)} \quad \Phi_x^{-1}(x^s) \text{ does not exist }, \quad S_x^{-1}(x^s) \text{ does exist}$$

and therefore even if $\Phi_x^{-1}(x^s)$ does not exist, the inverse $S_x^{-1}(x^s)$ does exist and we can still use a damped Newton's Method to find a zero of the equivalent system $S(x)=0$.

## 2.4 Derivation of the Equivalent System S(x).

Detection of the Singularity. If the damped Newton's Method applied to the original system $\Phi(x)$, generates points $x_i$ , $i=1,2,\ldots k$ which are being attracted to a singular point $x^s$, we know the stepsize $\beta$ becomes very small; therefore if we detect that $\beta$ becomes small in the bisection process, say smaller than $\beta<1/2^5$ , we say that the present point $x_k$ used in the algorithm is in the neighborhood of a singular point $x^s$.

Cancellation of the Singularity. Once we know the approximate position of the singular point, we define the equivalent system as follows

$$S(x) = \frac{\Phi(x)}{\{(x-x^m)^T(x-x^m)\}^{\eta}} \tag{16}$$

where $x^m$ denotes the position of a "movable pole" and $\eta$ denotes its strength We define initially $x^m = x_k$ as the pole position, but later on it will be moved, hence its name. The rules to move it and recompute its pole strength are disscused in the following section.

## 2.5 Calculation of the Movable Pole Strength $\eta$ .

This is easily done by the computer in the following automatic way;

Starting with $\eta = 0$ we compute the Jacobian

$$S_x(x) = \frac{1}{\{(x-x^m)^T(x-x^m)\}^\eta} \cdot \{\Phi_x(x) - 2\eta \frac{(x-x^m)\ \Phi^T(x)}{(x-x^m)^T(x-x^m)}\} \tag{17}$$

at a point $x=x^m + \varepsilon$ , where $\varepsilon$ is a small random number, used just to avoid a division by zero in the computer. A tentative displacement $\Delta x$ is computed from the system of equations

$$S_x^T(x)\ \Delta x = -\ S(x) \tag{18}$$

and a new tentative position is obtained from $\tilde{x} = x + \beta\ \Delta x$.
If the Performance Index $Q(x) = S^T(x)S(x)$ does not satisfy the descent property $Q(\tilde{x}) < Q(x)$, we increase $\eta$ to $\eta + \Delta\eta$ and repeat the above process. If the condition $Q(\tilde{x}) < Q(x)$ is satisfied we keep the present value of the pole strength $\eta$ and $\tilde{x}$ as the new position vector for the next iteration on the equivalent system $S(x)$.

2.6 Calculation of the Movable Pole Position $x^m$.

The cancellation of the singularity is very effective with $x^m = x_k$ and $\eta$ computed as in sec 2.5. A typical shape of $S(x)$ is shown in Fig. 6.a for $(x-x^m)^T(x-x^m) \leq 1$ , and in Fig. 6.b for $(x-x^m)^T(x-x^m) > 1$ . This last shape, pulse like and mostly flat, is produced by the scalar factor $1 / \{(x-x^m)^T(x-x^m)\}^\eta$ appearing both in the $S(x)$ and $S_x(x)$ equations, particularly if $\eta$ is large, because the scalar factor makes $S(x) \to 0$ and $S_x(x) \to 0$, except in a small neighborhood of $x^m$. In general this shape causes Newton's Method to slow down, requiring a large number of iterations to find the zero of $S(x)$.
To avoid this unfavorable shape, we shall move the movable pole, whenever the condition $(x-x^m)^T(x-x^m) > R$ is satisfied, ( say $R=1$). We will move the pole to the last acceptable position $x_k$, that is we set

$$x^m = x_k \quad , \eta = 0 \tag{19}$$

and the pole strength $\eta$ is recomputed for this new pole position as described in section 2.5 .

Summary of the Stabilized Newton's Method. A simplified flowchart of the method is given in Fig.7. We note a nice simple nested structure, unifying the structure of the Undamped, ( $\eta=0$, $\beta=1$ ) , Damped, ( $\eta=0$, $<\beta\leq 1$) , and the present Stabilized Newton's Methods, ($\eta \geq 0$, $<\beta\leq 1$).

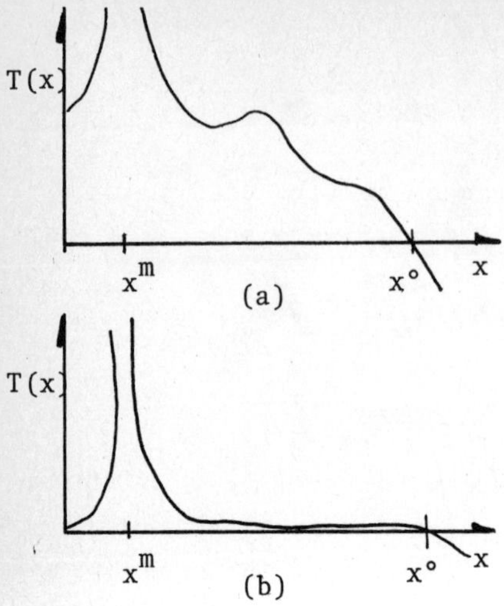


Fig. 6 Typical shapes of S(x).
(a). Favorable. $(x-x^m)^T(x-x^m) \leq 1$
(b). Unfavorable; $(x-x^m)^T(x-x^m) > 1$

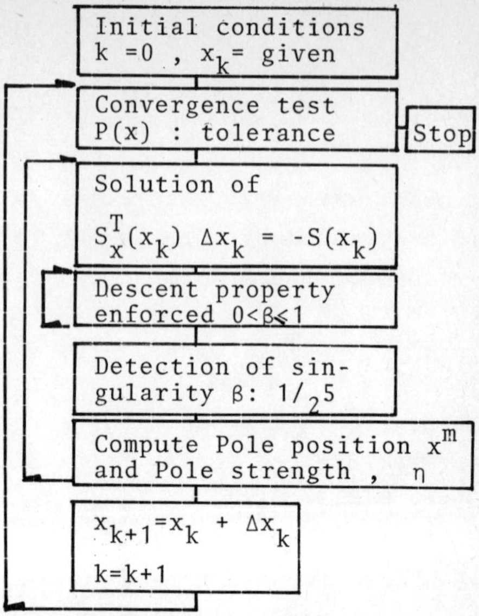


Fig. 7 Simplified Flowchart of the Stabilized Newton's Method, illustrating the nice, nested structure, unifying the Undamped, Damped and Stabilized Newton's Methods.

## 2.7 Numerical Examples.

A complete list of the examples can be found in Ref. 4. They vary from one equation to a system of five nonlinear equations.

Experimental conditions. The algorithm was programmed in FORTRAN IV in double precision, using a B-6800 computer. Three methods were compared Undamped, Damped and Stabilized Newton's Method. Each example was solved from 100 nominal points and the probability of succcess was computed as

$$p = \frac{N_S}{100} \tag{20}$$

where $N_S$ denotes the number of successful runs. If $t_i$ denotes the CPU computing time required for each nominal to converge, then the average computing time is given by

$$t_{av} = \frac{\sum_{i=1}^{N_S} t_i}{N_S} \tag{21}$$

Convergence was defined as $P(x) < 10^{-20}$. Nonconvergence was assumed if the number of iterations exceded a given limit, for each nominal (say 500 iter.)

| Ex. No. | UNDAMPED NEWTON | | DAMPED NEWTON | | STABILIZED NEWTON | |
|---|---|---|---|---|---|---|
| | p | $t_{av}$ | p | $t_{av}$ | p | $t_{av}$ |
| 1 | 0.78 | 1.29 | 0.72 | 0.18 | 1.00 | 0.48 |
| 2 | 0.67 | 0.37 | 0.98 | 0.10 | 1.00 | 0.22 |
| 3 | 0.81 | 0.47 | 1.00 | 0.08 | 1.00 | 0.16 |
| 4 | 0.56 | 0.41 | 0.53 | 0.69 | 0.81 | 6.60 |
| 5 | 0.42 | 1.46 | 0.79 | 0.27 | 1.00 | 0.26 |
| 6 | 0.67 | 0.41 | 0.94 | 0.28 | 1.00 | 0.56 |
| 7 | 0.98 | 0.92 | 0.84 | 0.23 | 1.00 | 1.08 |
| 8 | 0.44 | 0.07 | 0.78 | 0.33 | 0.99 | 0.84 |
| 9 | 0.49 | 0.24 | 0.50 | 0.15 | 0.96 | 1.92 |
| 10 | 0.41 | 3.97 | 0.24 | 0.92 | 0.63 | 3.80 |
| 11 | 0.76 | 10.20 | 0.21 | 0.96 | 0.97 | 7.47 |

Fig. 8 Numerical Results.
p = probability of success
$t_{av}$ = Average computing time (sec.)

## 2.8 Conclusions .

The numerical results are shown in Fig. 8.This results indicate that ;
a). The Stabilized Newton's Method has a higher probability of converging to the solution,than the Undamped and Damped Newton's Method.If $\tilde{p}$ denotes the average over the 11 examples ,the Undamped has $\tilde{p}$ =0.64, the Damped has $\tilde{p}$ = 0.68 and the Stabilized Newton Method has $\tilde{p}$ =0.91 .We can say with confidence, that it is a Robust Method.
b). The convergence radius has been greatly increased; the nominals are very far away from the solution with many singular points.
c) The computing effort increases,due to the stabilization procedure in each iteration,however this increment is worthwhile and not exagerated, if $\tilde{t}_{av}$ denotes the average over the 11 examples, the Undamped has $\tilde{t}_{av}$= 1.80 sec. , the Damped has $\tilde{t}_{av}$ = 0.38 sec., and the Stabilized has $\tilde{t}_{av}$= 2.12 sec.

## 2.9 Use of the Stabilized Method in the Tunnelling Algorithm.

In order to find the zero of the Tunnelling Function, using the concept of the movable pole described in the Stabilized Newton's Method, the following expression is employed, instead of Eq.4

$$T(x,\Gamma) = \frac{f(x) - f(x^*)}{\{(x-x^m)^T(x-x^m)\}^\eta \, .\{(x-x^*)^T(x-x^*)\}^\lambda} \tag{22}$$

where $\Gamma$ denotes th set of parameters $\Gamma=(x^*,\lambda,x^m,\eta)$. $x^m$ and $\lambda$ are updated at each iteration following the procedure for each parameter as indicated in the previous sections.

## 3. CONSTRAINED GLOBAL MINIMIZATION.

### 3.1 Statement of the Problem.

In this section the problem of finding the global minimum of f(x) subject to the constraints $g(x)\geqslant 0$ and $h(x)=0$ is considered,where f is a nonlinear scalar function, x is an n-vector, g and h are m and p non-linear functions defining the feasible set.Furthermore x is limited to the hypercube $A\leqslant x\leqslant B$ where A and B are prescribed n-vectors.

The feasible set can be a convex set or the constraints may defined several non-convex,non-conected feasible sets.Due to the importance of this topic, it is presented in a companion paper "Global Optimization of Functions Subject to Non-convex,Non-connected Regions." by Levy and Gomez.

## REFERENCES

1.A. V. Levy and A. Montalvo, " The Tunnelling Algorithm for the Global Minimization of Functions ", Dundee Biennal Conference on Numerical Analysis,Univ. of Dundee,Scotland,1977.

2.A.V. Levy and A. Montalvo,"Algoritmo de Tunelizacion para la Optimacion Global de Funciones", Comunicaciones Tecnicas, Serie Naranja,No. 204, IIMAS-UNAM,Mexico D.F.,1979.

3. A.V. Levy and A. Calderon, "A Robust Algorithm for Solving Systems of Non-linear Equations", Dundee Biennal Conference on Numerical Analysis, Univ. of Dundee,Scotland,1979.

4. A. Calderon,"Estabilizacion del Metodo de Newton para la Solucion de Sistemas de Ecuaciones No-Lineales" ,B. Sc. Thesis, Mathematics Dept. UNAM, Mexico D.F. ,1978.

5. A.V. Levy and A. Montalvo," A Modification to the Tunnelling Algorithm for Finding the Global Minima of an Arbitrary One Dimensional Function ", Comunicaciones Tecnicas,Serie Naranja,No. 240,IIMAS-UNAM,1980.

6. D. M. Himmelblau, "Applied Nonlinear Programing",Mc Graw-Hill Book Co. N.Y. ,1972.

## APENDIX : TEST PROBLEMS

Example 1.

$$f(x) = x^6 - 15x^4 + 27x^2 + 250$$

$$-4 \leqslant x \leqslant 4$$

Example 2.

$$f(x) = -\sum_{i=1}^{5} i\cos\{(i+1)x + i\}$$

$$-10 \leqslant x \leqslant 10$$

Example 3.

$$f(x,y) = \left(\sum_{i=1}^{5} i\cos\{(i+1)x+i\}\right)\left(\sum_{i=1}^{5} i\cos\{(i+1)y+i\}\right) +$$

$$\beta\{(x + 1.42513)^2 + (y + 0.80032)^2\}$$

$$-10 \leqslant x,y \leqslant 10 , \quad \beta = 0.$$

Example 4.

See Example 3., with $\beta = 0.5$

Example 5.

See Example 3. , with $\beta = 1.0$

Example 6.

$$f(x,y) = (4 - 2.1x^2 + x^4/3)x^2 + xy + (-4 + 4y^2)y^2$$

$$-3 \leqslant x \leqslant 3 , \qquad -2 \leqslant y \leqslant 2$$

Example 7

$$f(x) = \pi\{k\sin^2\pi y_1 + \sum_{i=1}^{n-1}\{(y_i - A)^2(1 + k\sin^2\pi y_{i+1})\} + (y_n - A)^2\}/n$$

with A=1 ,k=10 , n=2 and $y_i = 1 + (x_i - 1)/4$

$$-10 \leqslant x_i \leqslant 10, \ i = 1,2,\ldots,n$$

Examples 8,9,10,11,12, See Ex. 7 with n=3,4,5,8,10 ,respectively.

Example 13.

$$f(x) = k_1\sin^2\pi l_o x_1 + k_1\sum_{i=1}^{n-1}\{(x_i - A)^2(1 + k_o\sin^2\pi l_o x_{i+1})\}$$

$$+ k_1(x_n - A)^2(1 + k_o\sin^2\pi l_1 x_n)$$

$A = 1$, $l_o = 3$ y $l_1 = 2.$ $k_o = 1$, $k_1 = 0.1$ $n = 2$, $-10 \leqslant x_i \leqslant 10$

Examples 14,15, See Ex. 13 with $-10 \leqslant x_i \leqslant 10$ and n=3,4 respectively

Examples 16,17,18, See Ex. 13 with $-5 \leqslant x_i \leqslant 5$ and n=5,6,7 resp.

# THE TUNNELLING METHOD FOR SOLVING THE CONSTRAINED GLOBAL OPTIMIZATION PROBLEM WITH SEVERAL NON-CONNECTED FEASIBLE REGIONS

SUSANA GOMEZ & A.V. LEVY
IIMAS - UNAM
Apdo. Postal 20-726
01000 México, D.F. MEXICO

ABSTRACT.- The problem of finding the global minimum of f(x) subject to the constraints $g(x) \geq 0$ and $h(x)=0$ is considered, where the feasible region generated by the constraints could be non-convex, and could even be several nonconnected feasible regions.

An algorithm is developed which consists of two phases, a "minimization phase" which finds a local constrained minimum and a "tunnelling phase" whose starting point is the local minimum just found, and where a new feasible point is found which has a function value equal to or less than at the starting point and this point is taken as the starting point of the next minimization phase, thus the algorithm approaches the global minimum in an orderly fashion, tunnelling below irrelevant local constrained minima. When nonconnected feasible regions are present the tunnelling phase moves from the minimum of a feasible set to another nonconnected feasible set where the function has lower or equal value. It ignores completely all feasible sets where the function attains a higher value.

Numerical results are presented for 6 examples, for the tunnelling algorithm and a multiple random method. These results show that the method presented here is a <u>robust</u> method and its robustness increases with the density of local minima, the dimension, and the number of nonconnected feasible regions.

1. <u>STATEMENT OF THE PROBLEM</u>. We want to find the global minimum

$$\min f(x) = f(x^*_G) \tag{1}$$

$$\text{subject to} \qquad g(x) \geq 0 \ , \quad h(x)=0, \quad A \leq x \leq B$$

Here x is an n-vector, f is a nonlinear scalar function, g and h are i and d nonlinear vector functions, and we assume that f, g and $h \in C^2$, and $x^*_G$ a global minimizer exists.

1.1 <u>Active Constraints</u>. Because some of the constraint functions are inequalities we will use an active constraint strategy. We have then to define the set of active constraints: the equality constraints $h_j(x)=0$, $j=1,2,\ldots,$ are always active and the active inequality constraints are the ones that satisfy $g(x) \leqslant \varepsilon$, where $\varepsilon$ is a small positive number. Then the active constraints $\phi(x)_v$ are defined as

$$\phi(x)_v=\{h_j(x),\ j=1,2,\ldots d\ ;\ g(x)/g_j(x) \leqslant \varepsilon,\ j=d+1,\ldots v\} \qquad (2)$$

we will use, for simplicity, notation $\phi(x)_v$ to denote the active constraints at point x, where v is the number of active constraints.

Although we are using the concept of $\varepsilon$-feasibility in forming the set of active constraints for inequalities, we use through out this work a measure of the constraint violation defined by

$$P(x)_v = \phi^T(x)_v\phi(x)_v \qquad (3)$$

this performance index has continuous isolines even when a change of the active basis is produced. See Figure 1.

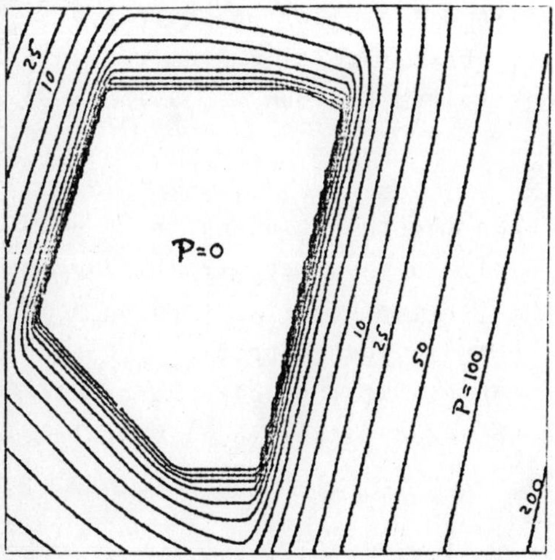

Figure 1

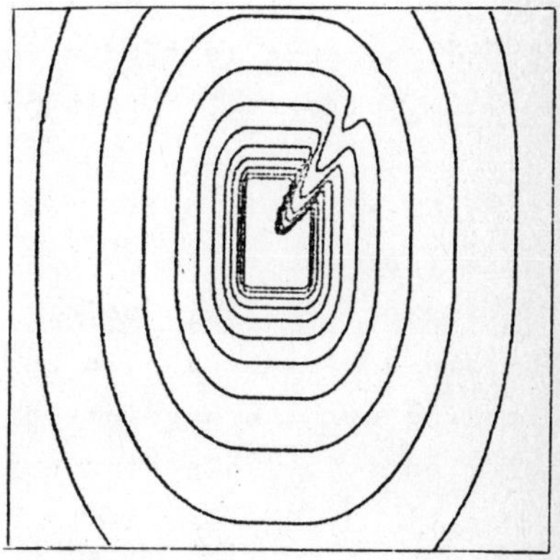
Figure 2

This property also holds when the feasible set is nonconvex, see figures 2 and 3 and even when there are nonconnected feasible sets, see figure 4.

NOTE: We use the same kind of performance index when testing the solution of the tunnelling phase. See section 1.6.

Figure 3

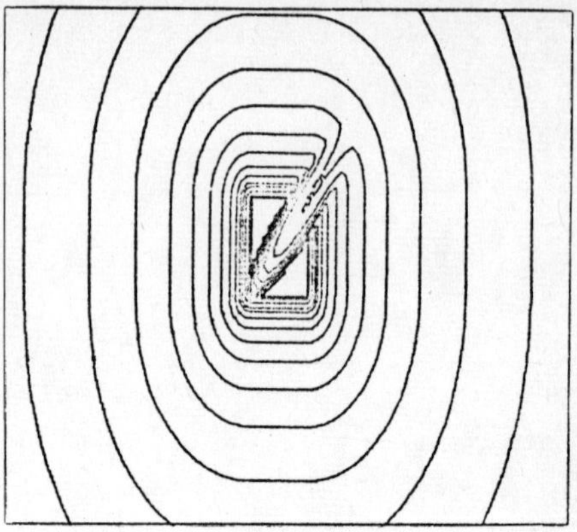

Figure 4

The algorithm will approach the global constrained minimum in a deterministic descent fashion, that is it will find a sequence of minimizer points $\{x_j^*\}$ such that $f(x_j^*) \leq f(x_{j-1}^*)$, $j=1,2,\ldots,G$ and constraint satisfaction $g(x_j^*) \geq 0$, $h(x_j^*)=0$, $A \leq x_j^* \leq B$, $j=1,2,\ldots,G$.

1.2 <u>Structure of the Algorithm</u>. In solving the constrained global problem we have also used the sequential structure described by Levy et.al., at the preceeding paper, that is a minimization phase and a tunnelling phase. In the minimization phase, starting from a feasible point $x_i^\circ$, we have to find a local constrained minimum $x_i^*$, using any algorithm suitable for this purpose.

However, it is often the case, when the feasible region is not convex, that there are points with local minima of constraint violation which are not feasible points, and then the methods for finding a local constrained minimum should be able to handle such a problem. One way of dealing with such points could be the use of the poles described in section 1.3. This strategy could be added to any local minimization method.

We have used here the sequential gradient restoration algorithm for inequality constraints due to A.V.Levy and S. Gomez ref.[2], which is a first order gradient projection method. We have added the strategy mentioned before. It is recommended to use a first order method when the density of local minima is high, although in the neighborhood of a local minimum a switch to a second order method is advisable. When there is evidence of few local minima a second order method for solving the local constrained problem should be used to improve the rate of convergence.

In the tunnelling phase, we have to find a feasible point $x^{\circ}_{i+1} \neq x^{*}_{i}$ satisfying

$$f(x^{\circ}_{i+1}) \leqslant f(x^{*}_{i}) \ , \ \phi(x^{\circ}_{i+1}) \geqslant 0 \quad , \quad A \leqslant x^{\circ}_{i+1} \leqslant B$$

If there is such a point, it will be in the zone of attraction of another local constrained minimum $x^{*}_{i+1}$ that will satisfy $f(x^{*}_{i+1}) \leqslant f(x^{*}_{i})$. The next minimization phase will get such a point $x^{*}_{i+1}$.

1.3 Classification of the Point $x^{*}_{i}$. In the tunnelling phase, in order to get $x^{\circ}_{i+1}$ we have to be able to move away from $x^{*}_{i}$ and in doing so we have first to classify $x^{*}_{i}$. There are three cases,

a) $x^{*}_{i}$ is an interior feasible point, that is $x^{*}_{i}$ is also the solution of the unconstrained problem

$$\min_{x} f(x) \tag{5}$$

Because it could be useful in another context we state that $x^{*}_{i}$ could be any stationary point of $f(x)$. i.e. $f_x(x^{*}_{i})=0$, and we are still able to move away from that point if needed.

b) $x^{*}_{i}$ is a boundary feasible point but also $x^{*}_{i}$ is the solution of the unconstrained problem Eq. (5).

Cases a) and b) can be treated in the same way.

c) $x^{*}_{i}$ is a boundary feasible point, but it is not the solution of problem Eq. (5). In this case the points x in a neighborhood of $x^{*}_{i}$ do not necessarity satisfy $f(x) < f(x^{*}_{i})$, and if they do, then $\phi(x) \ngeqslant 0$. This case must be treated in a different way because the functions have a different behaviour.

In all the cases we have to find a feasible point $x_i^\circ$ during the tunnelling phase; then the problem we have to solve here is to find a solution of the system of inequalities given by

$$t(x,\Gamma) = \begin{Bmatrix} T(x,\Gamma) \\ \\ -\phi(x)_v \end{Bmatrix} \leqslant 0 \qquad (6)$$

However the first inequality of the system, i.e. $T(x,\Gamma) \leqslant 0$ called the tunnelling function will be derived differently for cases a) and b) and for case c). See sections 1.4, 1.5.

In order to solve the system given by Eq. (6) any zero finding algorithm could be used, but we have used here the Stabilized Newton Method described by Levy et.al. at the preceding paper. That method deals with the problem of singular Jacobian when using Newton's method for solving a system of nonlinear equations. Here the system of inequalities $t(x,\Gamma)$ could have singular Jacobians specially due to the nature of the inequality $T(x,\Gamma)$ (tunnelling function). Then it is essential to use some method able to deal with singularities. The system we have to solve when using the Stabilized Newton Method will now be

$$t(x,\Gamma) = \frac{1}{[(x-x_m)^T(x-x_m)]^n} \begin{Bmatrix} T(x,\Gamma) \\ \\ -\phi(x)_v \end{Bmatrix} \leqslant 0 \qquad (7)$$

the computation of the pole strength n was described by Levy et.al. in section 2.5.

The position of the pole $x_m$, in cases a) and b) described above will be as described in section 2.6 by Levy et.al.. In case c) computation of $x_m$ will be described in section 1.5 where we will derive and explain the behaviour of the functions in the case when $x_i^*$ is not an unconstrained minimum of $f(x)$. Therefore the position of the pole $x_m$ will be chosen to get both, convergence to the solution of system (7) (in the presence of singular Jacobian) and to help to prevent going back to the local minimum just found.

We will now describe how the tunnelling function $T(x,\Gamma)$ is generated.

1.4 Derivation of the Function $T(x,\Gamma)$ for cases a) and b). Because $x_i^*$ is also a local minimizer of problem Eq. (5), in a neighborhood of $x_i^*$ there is no point x with $f(x)=f(x_i^*)$ and then we can destroy such a minimum using the same function used for unconstrained minimization, that is

$$T(x,\Gamma) = \frac{f(x)-f(x_i^*)}{\{(x-x_i^*)^T(x-x_i^*)\}^\lambda} \tag{8}$$

the deflating effect of this function was already explained by Levy et. al.. It generates a pole placed at $x_i^*$ which actually destroys the minimum at the point. The calculation of the pole strength $\lambda$ has been already described by Levy et.al.

1.5 Derivation of Function $T(x,\Gamma)$ for case c). Because $x_i^*$ is a constrained minimum, there is a region in the neighborhood of $x_i^*$ with points x arbitrarily close to $x_i^*$, satisfying $f(x) \leqslant f(x_i^*)$ and hence for $T(x,\Gamma)$ as in Eq. (6), $T(x,\Gamma) \leqslant 0$. These points are not feasible so that the system Eq. (8) is not satisfied, but attempts to reduce the infeasibility while keeping $T(x,\Gamma) \leqslant 0$ will drive the iterates back towards the constrained minimum $x_i^*$. However we can avoid going back to $x_i^*$ by placing a pole at $x_c$ between the iterates and $x_i^*$ with the needed intensity . Then at every point we have to recompute these parameters $x_c$, $\lambda$ (position and intensity of the pole), and the new tunnelling function will now be defined as

$$T(x,\Gamma) = \frac{f(x) - f(x_i^*)}{\{(x-x_c)^T(x-x_c)\}^\lambda} \tag{9}$$

$x_c$ will be a point on the line defined by x and $x_i^*$ so that $x_c = \beta x_i^* + (1-\beta)x$ where $\beta$ is such that $0 \leqslant \beta \leqslant 1$, and is the smaller $\beta$ such that $\|x-x_c\| < \tau_3 \leqslant 1$.

If we call $f_{xrad}$ the projection of $f_x(x)$ on the line $x-x_c$,

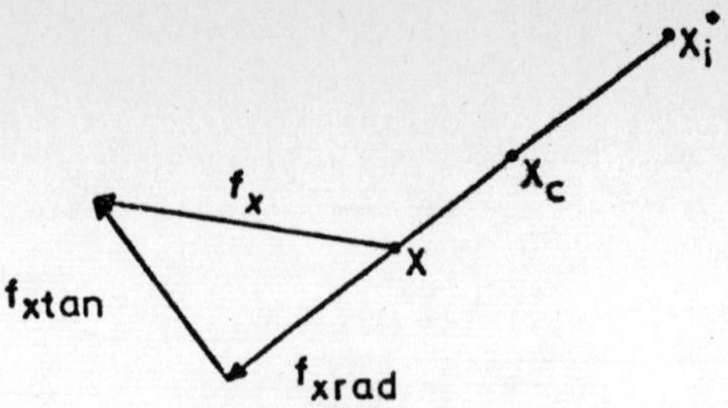


Figure 5

then from the expression of the gradient of $T(x,\Gamma)$ with respect to x, we can find explicitly $\lambda$, by impossing $\|f_{xrad}\| \geq \|x-x_c\|$ and then avoiding going back to $x_i^*$, $\lambda$ will then be computed by:

$$\lambda = \frac{f_x^T(x)\ (x-x_c)}{2|f(x)-f^*|} \tag{10}$$

$\lambda$ and $x_c$ must then be calculated at each iteration of the tunnelling phase. Once we have computed the tunnelling function $T(x,\Gamma)$ we will have to solve system Eq. (7) and in doing so it could be necessary to use the movable pole to destroy singularities of this system. But in this case we will take also advantages of these movable poles to enforce the poles at $x_c$, and accelerate convergece of the stabilized Newton Method to get the solution of system Eq. (4), at the same time. Then the position of these movable poles will now be calculated as follows: At the first iteration of the tunnelling phase take $x_m = x_i^*$. At subsequent iterations compute:

$$\|f_{xtan}\| = \|f_x(x) - \frac{|f_x^T(x)(x-x_c)|}{(x-x_c)^T(x-x_c)} \cdot \frac{(x-x_c)}{(x-x_c)^T(x-x_c)}\|$$

If $\|f_{xtan}\| < \varepsilon$, start with $\beta=1$, and using a bisection process on $\beta$ compute $x_m = \beta x_c + (1-\beta)x$ until condition $\|x-x_m\| \leq \tau_2$ holds.

If $\|f_{xtan}\| \geq \varepsilon$ compute $x_m = x + \tau_2 \frac{f_{xtan}}{\|f_{xtan}\|}$ where $\varepsilon$ is a small preselected number. This position $x_m$ decreases $f(x)$ as $t(x,\Gamma)$ approaches the solution.

In all cases a), b) or c) it could happen that $x_{i+1}^{\circ}$ is again a local minimum, and then at stage k of the algorithm we could have up to $\ell$ minima with the same value of the function $f(x_k^*) = f(x_i^*) = f^*$, $i=1,2,\ldots,\ell-1$.

To avoid going back to the minima found earlier, we keep the information, $x_i^*$, $\lambda_i$ of all these points, as well as $f^*$, and we now generate the function $T(x,\Gamma)$ as

$$T(x,\Gamma) = \frac{f(x) - f^*}{\prod_{i=1}^{\ell} \{(x-x_{ci})^T(x-x_{ci})\}^{\lambda i}} \qquad (11)$$

However as soon as a better value of $f^*$ is found, the stored information is removed, setting $\ell=1$.

1.6 Stopping Conditions. We define as an approximate solution of system, Eq. (7) if one of the following conditions is satisfied

$$Q(x_{i+1}^\circ) = t^T(x_{i+1}^\circ,\Gamma)t(x_{i+1}^\circ,\Gamma) \leqslant \varepsilon_1 \qquad (12)$$

or

$$f(x_{i+1}^\circ) \leqslant f(x_i^*) \quad \text{and} \quad \phi(x_{i+1}^\circ) \geqslant 0 \qquad (13)$$

The point $x_{i+1}^\circ$ which satisfies either (12) or (13) will now be the nominal point for the next minimization phase of the algorithm.

Of course the tests in (12) or (13) cannot be satisfied if (7) has no solution, which implies that $x_i^*$ is the global minimum. There is no clear test for the existence of a solution to (7), so the following stopping condition is used:

a) start tunnelling for 2n+5 nominal points, $x_j$, $x_j = x_i^* + \Delta x_j$, $j=1,2,..,2n+5$ where the $\Delta x_j$ are chosen in the following fashion:
   i) The first 2n points are: $\Delta x_j = \pm\tau_1 e_j$, $j=1,...,n$; here $e_j$ are the unitary vectors, and $|\tau_1| \ll 1$.
   ii) Then $\Delta x_j$, $j=2n+1,...,2n+5$ are random vectors such that $A \leqslant x_j \leqslant \beta$.

b) For each nominal point, $x_j$ allow 100 iterations for the stabilized Newton Method to find the solution of system, Eq. (7); call the number of iterations NIT and the number of starting points $x_j$, NSP. Then system Eq. (7) is assumed to have no solution if NSP $\geqslant$ 2n+5 NIT $\geqslant$ 100, thus $x_i^*$ is probably the global minimum, and the algorithm terminates.

2. <u>NONCONNECTED FEASIBLE REGIONS</u>. As we have solved the global constrained problem Eq. (1) by solving the system of inequalitites $t(x,\Gamma) \leqslant 0$, Eq. (7) and non feasibility is obviously allowed at intermediate points, it is then possible to tunnel from one feasible region to another.

Because these problems are the most difficult ones, we will give some numerical results obtained for problems with nonconnected feasible regions and we will compare them with results obtained when using a multiple random method (MRS). For both the tunnelling method and the MRS, we have used the same local minimization method mentioned in section 1.2. See ref. [2].

2.1 <u>Numerical Results</u>. The list of the examples with nonconnected feasible regions is given in the appendix. Numerical results for 12 examples with one nonconvex feasible region were given in Ref. [3]. The tunnelling algorithm was programmed in FORTRAN IV, in double precision. A B-6700 computer was used.

The Sequential Gradient-Restoration Algorithm for inequality constraints was used in the minimization phase. The step-size was computed using cubic interpolation, enforcing a local descent property until condition $-F_x^T(x+\alpha p,\lambda_v)p(x) \leqslant 10^{-4}\ p(x)^T p(\dot{x})$ is satisfied. The stopping conditions were $Q(x,\lambda_v) = F_x^T(x,\lambda_v)\ F_x(x,\lambda_v) < 10^{-9}$ where $F_x(x,\lambda_v)$ is the gradient of the lagrangian. In the tunnelling phase the Stabilized Newton Method was employed. The tunnelling phase was terminated as described in section 1.6.

We have used 20 nominal random points, for each problem, using the tunnelling algorithm and the Random Method MRS. Because of time limitations for each run, we had to give several seeds.

In Table No. 1 we report time (CPU,seconds), the number of function evaluations $f_{ev}$, and the number of constraint evaluations $\phi_{ev}$, required for these 20 nominal points.

We also report the efficiency of each method, that is, starting from $N_R$ starting points, the ratio of the number of global minima found $N_S$ to the total number of global minima which should be found for both methods, that is if the number of global minima of each example is $N_G$ then eff.=$N_S/N_R \cdot N_G$. For every nominal point we imposed a time limit which we took as 5 minutes CPU time. If the tunnel method did not get to the global minimum within this time we counted it as a failure.

In almost all the examples runned, the efficiency of the MRS method was very low compared with the efficiency gained using the tunnelling method, and hence we made further tests by letting the MRS method run the same time as used by the tunnelling method to run the 20 nominal points, rather than simply solving for 20 nominal points. We report these results in Table No. 1, using the notation $MRS^*$, giving the number of nominal points it generated and the efficiency of the method, that is the number of times it reached the global minimum over the number of starting points.

The tunnelling algorithm was shown to be much more efficient for all the examples. As described in the last section, we even let the MRS run many more than 20 nominal points (results with $MRS^*$), and we got no appreciable improvement.

For problems 4, 5 and 6 for which the feasible regions are marbles floating in a box in three dimensions, the tunnel did not get efficiency one due to the difficulty of the problem. In this case we only report the $MRS^*$ results for comparison. These results show that the tunnelling method is a robust method able to handle very difficult problems.

| Problem No. | 1 | | | 2 | | | 3 | | |
|---|---|---|---|---|---|---|---|---|---|
| | TUNN | MRS | $MRS^*$ | TUNN | MRS | $MRS^*$ | TUNN | MRS | $MRS^*$ |
| t | 233 | 46 | 233 | 545 | 27.1 | 565 | 946 | 135.9 | 946 |
| fev. | 3340 | 711 | 3458 | 12495 | 1490 | 25448 | 21051 | 7542 | 53541 |
| $\phi$ev. | 5259 | 1127 | 5740 | 16704 | 1577 | 28030 | 37455 | 8242 | 58816 |
| efficiency | 1 | 0.3 | 0.28 | 1 | 0.05 | 0.055 | 1 | 0.35 | 0.33 |
| No. Nominals | 20 | 20 | 106 | 20 | 20 | 363 | 20 | 20 | 164 |

| Problem No. | 4 | | | 5 | | | 6 | | |
|---|---|---|---|---|---|---|---|---|---|
| | TUNN | MRS | MRS* | TUNN | MRS | MRS* | TUNN | MRS | MRS* |
| t | 118 | 54.4 | 297 | 384 | 37.6 | 458 | 558 | | 762 |
| fev. | 1563 | 694 | 3613 | 9464 | 2060 | 20920 | 10477 | | 10857 |
| $\phi$ev. | 2205 | 1137 | 6860 | 12767 | 2263 | 24552 | 12923 | | 12264 |
| efficiency | 0.9 | 0.6 | 0.66 | 0.9 | 0.15 | 0.05 | 0.5 | | 0.089 |
| No. Nominals | 20 | 20 | 180 | 20 | 20 | 350 | 20 | | 101 |

TABLE No. 1

CONCLUSIONS. The method developed for finding a constrained global minimum is termed the "constrained tunnelling method". It too consists of two phases, a minimization phase in which a local minimum is found, and a tunnelling phase in which a new feasible point in a "better valley" is sought. It has the following properties:

a) It tunnels below irrelevant local minima, which is an important feature when we are dealing with problems with a high density of local minima.

b) It has a global descent property.

c) It could be linked as a subroutine to any local minimization method available.

d) It allows non-feasibility at the intermediate points. Problems with nonconnected feasible regions can therefore be solved.

e) Although the tunnelling phase essentially solves a system of equations based on the current active set to find a new feasible point with the same objective function value, any feasible point with the same or a lower objective function value is accepted (allowing inequalities $T(x,\Gamma) \leqslant 0$, $\phi(x) \geqslant 0$). This increases the flexibility of the method, specially when the local minimum occurs in the interior of the feasible region.

f) The method turns out to be robust, and relatively insensitive to the values of the various user-defined parameters.

g) Above all, the method has a high probability of success in finding the global minimum, and its performance in this respect was markedly superior to the "Multistart Random Search" method on a wide range of problems.

SOME COMMENTS ON APPLICATIONS OF THE METHOD. There is increasing interest in the global optimization problem, and growing evidence that many problems of practical interest exhibit multiple local minima. This seems to be particularly true in the area of optimal process design and process synthesis, where there are significant potential gains. Generally these are constrained minimization problems with a non-convex feasible region, and the objective function, representing cost, is often also non-convex. The method presented here can then be used to solve such problems.

Because we are able to solve problems with nonconnected feasible regions, it was suggested to us that it might be possible to solve integer programming problems. Every integer, within a specific precision, becomes a feasible region and we have already successfully solved some problems using linear and quadratic objective functions. The results depend on the specified precision, but when we are away from the solution we can take large feasible regions, and once the tunnel phase detects that we have the required integer, we can increase precision. We are now trying to solve larger and more complicated problems.

In addition to these direct applications of global minimization, some of the ideas are also useful in a more general context. Levy and Segura have shown how pole placement can be used to deal with singularity of the Jacobian in solving a system of nonlinear equations. Earlier in this paper we have mentioned the use of pole placement to deal with a local minimum of constraint violations in connection with nonlinear programming algorithms in general. Finally, the tunnelling phase itself can be used to improve tests for termination in such algorithms. For example, in practical problems derivatives are often approximated using finite differences, and the error introduced may then give conflicting evidence between function and "gradient" evaluations, the latter indicating a possible descent direction which is not confirmed by subsequent function evaluations along that direction. The tunnelling phase can be used to escape from such a region, or confirm that it is indeed the minimum. Overall it can reasonably be claimed that the tunnelling method makes constrained global minimization a practical possibility, and thereby opens up a range of new application possibilities.

TEST PROBLEMS.

Problem No. 1.- $\min f(x)=-x\ -x$

Subject to $\{(x_1-1)^2+(x_2-1)\}\{\frac{1}{2a^1}-\frac{1}{2b^2}\}+(x_1-1)(x_2-1)\{\frac{1}{a^1}-\frac{1}{b^2}\}-1\geqslant 0$

$1-x_1\geqslant 0,\quad 1-x_2\geqslant 0,\quad x_1,x_2\geqslant 0$ with $a=2$, $b=0.25$

The feasible regions are shown in figure 6. This problem has two global minima.

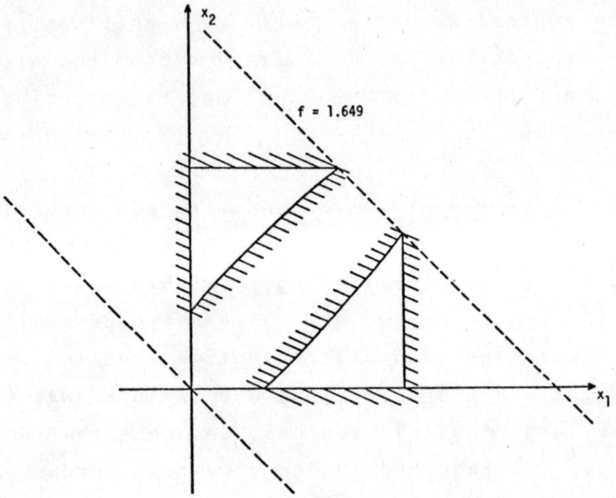


Figure 6

Problem No. 2.- $\min f(x)=0.1(x_1^2+x_2^2)$

subject to $\sin(4\pi x_1)-2\sin^2(2\pi x_2)\geqslant 0$

$-1\leqslant x_1,\ x_2\leqslant 1$

Problem No. 3.- $\min f(x)=\{4-2.1x_1^2+\frac{x_1^4}{3}\}x_1^2+x_1x_2+(-4+4x_2^2)x_2^2$

subject to same constraints that problem No. 2.

Problem No. 4.- $\min f(x)=-x_1-x_2+x_3$

subject to $\sin(4\pi x_1)-2\sin^2(2\pi x_2)-2\sin^2(2\pi x_3)\geqslant 0$

$-5\leqslant x_1,\ x_2\leqslant 5$

Problem No. 5.- $\min f(x)=x_1^2+\frac{1}{2}x_2^2+\frac{1}{4}x_3^2$

subject to same constraints that problem No. 4.

Problem No. 6.- $\min f(x)=\{4-2.1x_1^2+\frac{x^4}{3}\}\,x_1^2+x_1x_2+(-4+4x_2^2)x_2^2+x_3^2$

subject to same constraints that problem No. 4.

The feasible regions are shown in figure 7. Examples 2 to 6 have only one global minimum.

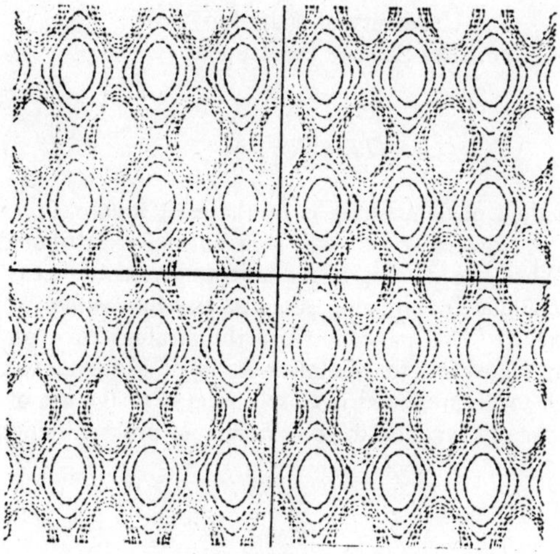

Figure 7


REFERENCES.

1.- A. V. Levy and A. Calderón, "A robust algorithm for solving systems of non-linear equations", Dundee Biennal Conference on Numerical Analysis, Univ. of Dundee, Scotland, 1979.

2.- A. V. Levy and Susana Gómez, "Sequential gradient restoration algorithm for the optimization of nonlinear constrained functions", Comunicaciones Técnicas, Serie Naranja, No. 189, IIMAS-UNAM, 1979.

3.- A. V. Levy and Susana Gómez, "The tunnelling algorithm for the global optimization of constrained functions", Comunicaciones Técnicas, Serie Naranja; No. 231, IIMAS-UNAM, 1980.

# An Approach to Nonlinear $l_1$ Data Fitting

*Richard H. Bartels*

The University of Waterloo
Computer Science Department

*Andrew R. Conn*

The University of Waterloo
Computer Science Department

*ABSTRACT*

The traditional method of data fitting is by the least squares ($l_2$) technique. When the data is good -- reasonably accurate with normally distributed errors -- this method is ideal. When the data is bad -- contaminated by occasional wild values -- then the $l_1$ technique (minimizing sums of absolute values of residuals) has much to recommend it. This paper surveys the strategy of a globally and superlinearly convergent algorithm to minimize sums of absolute values of $C^2$ functions. The approach to be presented is closely related to the use of a certain, piecewise differentiable penalty function to solve nonlinear programming problems.

## 1. Introduction

In this paper we outline the strategy of a method for solving the problem

$$\underset{x}{\text{minimize}}\ F(x) = \sum_{i=1}^{m} |f_i(x)| \tag{1.1}$$

where $x \in \mathbf{R}^n$ and $f_i \in C^2$ for all $i \in \{1, \ldots, m\}$. Such problems arise, for example, in data fitting contexts, where the functions $f_i$ are defined by

$$f_i(x) = h(t_i;x) - y_i \ .$$

Here the $y_i$ are considered to be observations of the functional $h(t,x)$ at $t_i$. If $h$ is linear in $x$; that is, $h(t_i;x) = h_i^T x$, then (1.1) is equivalent to a specialized linear programming problem

$$\underset{u,v,x}{\text{minimize}} \sum_{i=1}^{m} (u_i + v_i) \tag{1.2}$$

$$\text{subject to } h_i^T x + u_i - v_i = y_i \ ; \ i = 1, \ldots, m$$

$$\text{and} \quad u_i \geqslant 0,\ v_i \geqslant 0, \ ; \quad i = 1, \ldots, m$$

but with $x$ unrestricted.

The conventional method of fitting data $y$ to a functional form $h$ would replace the absolute value in (1.1) by the square, giving an $l_2$ estimation of $x$. Such estimates are very sensitive to the presence of outliers -- occasional observations $y_i$ which are wildly out of line with the rest. The $l_1$ estimate of $x$ has a certain capacity to ignore outliers. A simple linear example will suffice to illustrate.

Let

$$h(t;x) = x_1 + tx_2 \ ,$$

and consider the data

| t | y |
|---|---|
| 1 | 0.75 |
| 2 | 2.00 |
| 3 | 3.00 |
| 4 | 4.25 |
| 5 | 4.75 |
| 6 | 6.50 |
| 7 | 7.25 |
| 8 | 0.00 |

which was obtained, very roughly, from the form $h$ with $x_1 = 0$, $x_2 = 1$. Note that the $y$ value corresponding to $t = 8$ is wild. (Perhaps an 8 in the first position was misread as a zero when the data was collected or transcribed.) The $l_1$ estimation of $x$ for this data is $\bar{x}_1 = -0.1875$, $\bar{x}_2 = +1.0625$, which is not unreasonable, considering the accuracy of the good portion of the data. The $l_2$ estimation of $x$ for this data is $\hat{x}_1 = +1.848$, $\hat{x}_2 = +0.381$. A plot of the data and the two lines

$$\bar{y} = \bar{x}_1 + t\bar{x}_2 \qquad (l_1\text{line})$$

$$\hat{y} = \hat{x}_1 + t\hat{x}_2 \qquad (l_2\text{line})$$

reveals that the $l_1$ line accurately reproduces the sense of the first seven data values, ignoring the eighth completely. The $l_2$ line is as strongly influenced by the eighth data value as it is by all of the first seven put together. Consequently, the $l_2$ line shows no reasonable agreement with any portion of the data.

It is interesting to note that the $l_1$ line for the above example interpolates two of the data points, the third and the seventh. That is

$$f_3(\bar{x}) = h(t_3;\bar{x}) - y_3 = 0$$

$$f_7(\bar{x}) = h(t_7;\bar{x}) - y_7 = 0 \; .$$

This results from the fact that $\bar{x}$ is determined from the linear program (1.2), in which the components of $x$ are unrestricted, implying that these two components must be basic at an optimum. And the structure of (1.2) in turn implies that $u$, $v$ pairs associated with two equality constraints must be nonbasic at an optimum; i.e. the corresponding two deviations must be zero. More generally, if $x \in \mathbf{R}^n$, one can expect to have at least $n$ of the data points interpolated.

The algorithm surveyed in this paper is a natural extension of the algorithm in [1] to nonlinear problems, using the material of [3,4] as a theoretical foundation. As in [1], we exploit the interpolation feature.

For the sake of simplicity, no constraints appear in (1.1). The discussion in [1] however, where linear constraints for the linear version of (1.1) were incorporated by adding penalty terms to $F$ which could be handled by the algorithm in a natural way, generalizes directly to nonlinear problems with linear and nonlinear constraints. Our current software, in fact, solves the problem

$$\underset{x}{\text{minimize}} \quad \sum_{i=1}^{m} |f_i(x)|$$

$$\text{subject to} \quad f_j(x) \geqslant 0, \quad \text{for } j = m+1, \ldots, m+l$$

$$\text{and} \quad f_k(x) = 0, \quad \text{for } k = m+l+1, \ldots, m+l+s$$

The exposition is further simplified by assuming the availability of exact first and second derivatives rather than considering gradient differencing and/or quasi-Newton methods, though both can be employed in obvious ways.

## 2. Notation, Definitions, Assumptions

All norms ($\|\cdot\|$) used below are Euclidian norms.

We will organize our thoughts around the set (possibly vacuous) of those $f_i$ in (1.1) which will be zero at a local minimum of $F$. For a given $x$ and $\epsilon \geqslant 0$ let

$$\mathbf{A}(x,\epsilon) = \{j : |f_j(x)| \leqslant \epsilon \textbf{ and } 1 \leqslant j \leqslant m\}$$

denote the set of $\epsilon$*−active indices* at $x$. Let

$$\mathbf{I}(x,\epsilon) = \{1, \ldots, m\} - \mathbf{A}(x,\epsilon)$$

represent the $\epsilon$*−inactive indices* and

$$\sigma_i(x) = \text{sgn}[f_i(x)]; \quad i \in \mathbf{I}(x,\epsilon) .$$

Let

$$\phi(x,\epsilon) = \sum_{i \in \mathbf{I}(x,\epsilon)} \sigma_i(x) f_i(x) ,$$

and note that

$$\phi(x,0) = \sum_{i \in \mathbf{I}(x,0)} \sigma_i(x) f_i(x) = \sum_{i \in \mathbf{I}(x,0)} |f_i(x)| = \sum_{i=1}^{m} |f_i(x)| = F(x)$$

for any $x$. It is easily seen, moreover, that for any designated point $\hat{x}$ there is an $\hat{\epsilon} > 0$ such that for all $0 \leqslant \epsilon \leqslant \hat{\epsilon}$

$$\mathbf{A}(\hat{x},\epsilon) = \mathbf{A}(\hat{x},0)$$

and

$$\phi(\hat{x},\epsilon) = F(\hat{x}) .$$

With $\epsilon$ fixed, there is a neighborhood $\hat{N}$ of $\hat{x}$ depending upon $\epsilon$, for which $x \in \hat{N}$ implies

$$\mathbf{A}(x,\epsilon) = \mathbf{A}(\hat{x},0) ;$$

and also given any $\delta > 0$ there is a neighborhood $\hat{N}_\delta \subseteq \hat{N}$ of $\hat{x}$ within which

$$|\phi(x,\epsilon) - F(x)| < \delta .$$

Thus, we approach the task of solving (1.1) from the point of view of selecting a trial active index set $\mathbf{A}$ and its complimentary index set $\mathbf{I}$ suggested locally by $x$ and some activity tolerance $\epsilon$. (Usually $\mathbf{A} = \mathbf{A}(x,\epsilon)$, but we have sometimes found it useful to consider $\mathbf{A} \subseteq \mathbf{A}(x,\epsilon)$, and so we will distinguish the $\mathbf{A}$ of our choice from $\mathbf{A}(x,\epsilon)$ in what follows.) Then, with $\mathbf{A}$ chosen, we begin solving

$$\underset{x}{\text{minimize}}\ \phi(x) = \sum_{i \in \mathbf{I}} \sigma_i(x) f_i(x) \tag{2.1}$$

$$\text{subject to } f_j(x) = 0; \ j \in \mathbf{A}$$

until we are sufficiently close to a minimum of $F$ to satisfy a convergence criterion, or until it is evident that we are dealing with the wrong selection of $\mathbf{A}$ in which case (2.1) is redefined (by redefining $\mathbf{A}$) for further steps of the algorithm. The approach to solving problems of the form (2.1) will be the one taken in [3,4], for which it can be shown, under reasonable assumptions, that the sequence of function values $F$ which are generated decrease monotonically to a locally minimal value.

For ease of discussion we will begin with the following

**Assumption**

1. The vectors $\nabla f_j(x)$, $j \in \mathbf{A}$ are linearly independent for each $x, \mathbf{A}$ we choose to consider. (See section 5 for further comments on this.)

We summarize the applicable results and defer their full development to a later paper.

**Definition:** A point $\bar{x}$ is a *stationary point* of $F$, if there exist scalars $u_j$, $j \in \mathbf{A}(\bar{x}, 0)$, such that

$$\sum_{i \in \mathbf{I}(\bar{x},0)} \sigma_i(\bar{x}) \nabla f_i(\bar{x}) = \sum_{j \in \mathbf{A}(\bar{x},0)} u_j \nabla f_j(\bar{x}) \ .$$

**Definition:** A stationary point $\bar{x}$ of $F$ is a *first-order point* of $F$, if

$$-1 \leqslant u_j \leqslant +1; \ j \in \mathbf{A}(\bar{x}, 0) \ . \tag{2.3}$$

**Definition:** A first-order point $\bar{x}$ of $F$ is a *strict second-order point* of $F$, if

$$-1 < u_j < +1, \ \ j \in \mathbf{A}(\bar{x}, 0) \ ,$$

and for each nonzero vector $p \in \mathbf{R}^n$ satisfying

$$p^T \nabla f_j(\bar{x}) = 0, \ \ j \in \mathbf{A}(\bar{x}, 0) \ , \tag{2.4}$$

it follows that

$$p^T B(\bar{x}) p > 0 \tag{2.5}$$

where

$$B(\bar{x}) = \sum_{i \in \mathbf{I}(\bar{x},0)} \sigma_i(\bar{x}) \nabla^2 f_i(\bar{x}) - \sum_{j \in \mathbf{A}(\bar{x},0)} u_j \nabla^2 f_j(\bar{x}) \ . \tag{2.6}$$

The following are corollaries of results from [2].

**Theorem 1:** A necessary condition for the point $\bar{x}$ to be a local minimizer for $F$ is that $\bar{x}$ be a first-order point for $F$.

**Theorem 2:** A sufficient condition for $\bar{x}$ to be a strong local minimizer for $F$ is that $\bar{x}$ be a strict second-order point for $F$.

If $\mathbf{A}$ is chosen as $\mathbf{A}(\bar{x}, 0)$ in (2.1), then the following is a standard result.

**Theorem 3:** A necessary condition for $\bar{x}$ to be an optimal point for (2.1) is that $\bar{x}$ be a stationary point for $F$.

The indication which we may take that our choice of (2.1) -- that is, our choice of $\mathbf{A}$ -- is an incorrect representation of (1.1) is that in proceeding toward an optimum of (2.1) we appear to be nearing a stationary point of $F$ which is not a first-order point of $F$. On such occasions we change (2.1) by *dropping* an index $j_0$ from $\mathbf{A}$ as suggested by the following sequence of definitions and results.

**Definition:** The values $u_j = u_j(x)$, $j \in \mathbf{A}$ defined from (2.1) for a given $x \in \mathbf{R}^n$ by the solution to the least squares problem

$$\underset{u}{\text{minimize}} \ \| \nabla f_{\mathbf{A}}(x) u - \nabla \phi \| \ , \tag{2.7}$$

where

$$\nabla \phi(x) = \sum_{i \in \mathbf{I}} \sigma_i \nabla f_i(x) \tag{2.8}$$

and $\nabla f_{\mathbf{A}}(x)$ is a matrix with columns $\nabla f_j(x)$, $j \in \mathbf{A}$, are the *first-order multiplier estimates* at $x$.

**Definition:** Any vector $p \in \mathbf{R}^n$ defines a *descent direction* for $F$ at a point $x$, if $F(x+\alpha p) < F(x)$ for all $\alpha > 0$ in some neighborhood of 0.

It follows easily from the material in [1] that, if $\bar{x}$ is a stationary point of (2.1) which is not a first-order point for $F$ and $u_j = u_j(x)$, $j \in \mathbf{A}$ are the associated first-order multiplier estimates, then the vector $d$ given by the solution to the linear system

$$\begin{aligned} \nabla f_{j_0}(\bar{x})^T d &= -\text{sgn}(u_{j_0}) \\ \nabla f_j(\bar{x})^T d &= 0, \quad j \in \mathbf{A} - \{j_0\}, \end{aligned} \tag{2.9}$$

where $j_0 \in \mathbf{A}$ is some index for which $|u_{j_0}| > 1$, will define a descent direction for $F$ at $\bar{x}$. This result can be extended for the nonlinear problem under discussion to the following.

**Theorem 4:** If $\bar{x}$ is a stationary point for (2.1) which is not a first-order point for $F$, if $x$ is close enough to $\bar{x}$ in the sense that $\mathbf{A}(x,\epsilon) = \mathbf{A}(\bar{x},0)$, and if $\mathbf{A}$ in (2.1) coincides with $\mathbf{A}(\bar{x},0)$, if $\sigma_i(x) = \text{sgn}[f_i(x)] = \text{sgn}[f_i(\bar{x})]$ for all $i \in \mathbf{A}$ and if $u_j$, $j \in \mathbf{A}$ are the first-order multiplier estimates at $x$, then the vector $d \in \mathbf{R}^n$ determined as in (2.9) with $\bar{x}$ replaced by $x$ will define a descent direction for $F$.

**Definition:** The step direction $d$ defined above will be called the *dropping direction*

The algorithm for (1.1) to be described uses the dropping direction whenever the selected version of (2.1) (i.e. the choice of **A**) is not an appropriate model for (1.1). Otherwise one or both of a pair of directions based directly upon (2.1) is used.

**Definition:** A *horizontal direction* at $x$ for (2.1) is a vector $h \in \mathbf{R}^n$ which solves the following equality-constrained quadratic programming problem

$$\underset{h}{\text{minimize}}\ \tfrac{1}{2} h^T Q h + h^T \nabla \phi(x) \tag{2.10}$$

$$\text{subject to } \nabla f_{\mathbf{A}}(x)^T j = 0$$

for a positive definite matrix $Q$. During the course of the algorithm two choices of $Q$ are considered:

$$Q = \nabla^2 \phi(x) + D \tag{2.11}$$

for a diagonal matrix $D$ with nonnegative diagonal entries (possibly zero) as needed to ensure positive definiteness, and (as suggested by (2.6))

$$Q = B(x) = \nabla^2 \phi(x) - \sum_{j \in \mathbf{A}} u_j \nabla^2 f_j(x), \tag{2.12}$$

where $u_j$, $j \in \mathbf{A}$ are the first-order multiplier estimates at $x$. If it is necessary to distinguish which choice of $Q$ is used to produce $h$, we will use $h^{\circ}$ to indicate that (2.11) was used and $h^{\circ\circ}$ to indicate that (2.12) was used.

**Definition:** Let $\tilde{x}$ be a designated "reference point". A *vertical direction* at $x$ referenced to $\tilde{x}$ is a vector $v \in \mathbf{R}^n$ which solves the following least squares problem

$$\underset{v}{\text{minimize}}\ \| \nabla f_{\mathbf{A}}(\tilde{x})^T v + f_{\mathbf{A}}(x) \|, \tag{2.13}$$

where $f_{\mathbf{A}}(x)$, $j \in \mathbf{A}$ (NB: $x$ **not** $\tilde{x}$) denotes the vector of function values $f_j(x)$, arranged consistently with the columns of $\nabla f_{\mathbf{A}}(\tilde{x})$.

During the course of our algorithm a line search will be used for dropping directions and for some horizontal directions. Whenever $x + \alpha^* d$ or $x + \alpha^* h$ is written for any point $x$, it will be assumed that $\alpha^* > 0$ has been chosen to provide *sufficient decrease* in the sense that

$$F(x + \alpha^* d) < F(x) - \eta [d^T \nabla \phi(x)]^2 \tag{2.14}$$

for some chosen tolerance $\eta > 0$, and similarly for $h$.

**Assumptions** (continued)

2. $f_i \in C^2$ for all $i$.
3. Any points $x$ to be considered are confined to a compact set **S**.
4. (2.14) holds with respect to each $\alpha^*$ chosen.
5. There are only finitely many stationary points $\bar{x}$ in **S**.
6. All first-order points of $F$ in **S** are strict second-order points.
7. There exist numbers $U \geqslant L > 0$, $\pi > 0$ such that

   $$L \| y \|^2 \leqslant y^T B(x_k) y \leqslant U \| y \|^2$$

   for any $y$ satisfying $y^T \nabla f_j(x_k) = 0$, all $j \in \mathbf{A}$, whenever $\| x_k - \bar{x} \| < \pi$ for any first-order point $\bar{x}$ in **S**.

Under the above we have the following results more or less directly from [3,4].

**Theorem 5:** Let $\mathbf{A} = \mathbf{A}(\bar{x}, 0)$ define the problem (2.1), and assume that $\bar{x}$ is a first-order point for $F$. Then there is a neighborhood of $\bar{x}$ such that, for all $\tilde{x}$, $x$ in that neighborhood,

$$F(x + v) < F(x) \ , \tag{2.15}$$

where $v$ is the vertical direction at $x$ referenced to $\tilde{x}$.

**Theorem 6:** Let $\bar{x}$ be a stationary point for (2.1), with $\mathbf{A} = \mathbf{A}(\bar{x}, 0)$. Assume that $\bar{x}$ is also a strict second-order point for $F$. Then there is a neighborhood of $\bar{x}$ such that, for any $x$ in that neighborhood with

$$F(x + h^{\circ\circ} + v) < F(x) \tag{2.16}$$

where $v$ is the vertical direction at $x + h^{\circ\circ}$ referenced to x.

**Theorem 7:** Given any instance of (2.1) and any $x$, the horizontal directions $h^{\circ}$ , $h^{\circ\circ}$ as given in (2.11), (2.12) respectively are descent directions for $F$.

**Definition:** For any designated point $x$, and any chosen $d$, $h \in \{h^{\circ}, h^{\circ\circ}\}$, $v$ consistent with the above, the transitions

$$\begin{aligned} x &\to x + \alpha^* d \\ x &\to x + \alpha^* h \\ x &\to x + v \\ x &\to x + h^{\circ\circ} + v \end{aligned} \tag{2.17}$$

will be called respectively a *dropping step*, a *horizontal step*, a *vertical step* and a *Newton step* taken from the point $x$.

We are now prepared to outline our algorithm and its convergence properties.

## 3. Algorithm: Strategy and Convergence

Assume that **A** has been chosen and we are considering the resulting problem (2.1) at some point $\hat{x}$. We can be either (a) very close to a stationary point of (2.1) which, by virtue of having correctly identified **A**, is a first-order point of $F$, (b) nowhere near a first-order point of $F$ and trying to move closer to a stationary point of (2.1) as a means of decreasing the value of $F$, (c) somewhat near a stationary point of (2.1) and interested in testing whether **A** is corectly chosen. Thus, we associate three regions with each stationary point $\bar{x}$ of (2.1). $R_1$ consists of those points $x$ which, loosely stated, are so distant from $\bar{x}$ that (2.2) is far from satisfied. In such circumstances the multiplier estimates $u_j$ will not be required, which is fortunate since they can be expected to be quite unreliable. $R_2$ consists of those points $x$ which are closer to $\bar{x}$ in the sense that (2.2) is rather well satisfied, and the associated first-order multiplier estimates clearly indicate the character of the true multipliers. $R_3$ consists of those points which are within the intersection of the neighborhoods mentioned by theorems 5 and 6 and so close to $\bar{x}$ that (2.2) is very well satisfied.

To be more specific, let $\tau_{12} > \tau_{23} > 0$ be given. Then

$$R_1 = \{x : \| \nabla f_{\mathbf{A}}(x)u(x) - \nabla\phi(x)\| \geqslant \tau_{12}\}$$

$$R_2 = \{x : x \notin R_1 \bigcup R_3 \textbf{ and } \| \nabla f_{\mathbf{A}}(x)u(x) - \nabla\phi(x)\| < \tau_{12}\}$$

$$R_3 = \{x : x \in \{\text{nbd. of Thm. 5}\} \bigcap \{\text{nbd. of Thm. 6}\} \textbf{ and } \| \nabla f_{\mathbf{A}}(x)u(x) - \nabla\phi(x)\| < \tau_{23}\}$$

In practice we can never determine exactly into which region our given point $\hat{x}$ falls, but this is inessential to the success of the algorithm. The global decrease of $F$ to a locally minimal value is guaranteed under reasonable assumptions, and these regions are used only to govern our strategy of approach to the corresponding minimizer $\bar{x}$. Failure to determine the regions optimally, at worst, defers the time after which superlinearity of the convergence rate sets in. Our desire is to avoid estimating multipliers and confine ourselves to horizontal steps (viz. (2.17)) in $R_1$, to estimate multipliers and determine whether a dropping step is necessary in $R_2$, to use vertical steps as well as horizontal steps in $R_2$ if dropping is not called for, and finally to switch to Newton steps for fast convergence to $\bar{x}$ in $R_3$.

We base our assumption about which region contains a given point $\hat{x}$ largely upon the magnitude of $\| \nabla f_{\mathbf{A}}(\hat{x})u - \nabla\phi(\hat{x})\|$, and we back up this assumption by verification tests which we apply to the dropping direction, to the vertical step and to the Newton step as appropriate.

With this general introduction, the algorithm is most easily described in terms of a table. The tolerance parameters $\tau_{12} > \tau_{23} > 0$ have been introduced above. Three more positive parameters $\delta_1$, $\delta_2$ and $\delta_3$ are needed for determining whether to reject respectively a dropping step direction, a vertical step or a Newton step on the grounds that they will not yield acceptable decrease in $F$. Each iteration of the algorithm starts with a current point $\hat{x}$, chooses $\mathbf{A}$ anew as $\mathbf{A}(\hat{x}, \epsilon)$, and uses $\hat{x}$, $\mathbf{A}$ to test a condition (column 1 in the table below) which serves to pick out a table row. Each row is associated with an assumption about the $R$-region in which $\hat{x}$ is located (column 2). A verification of this assumption may be carried out by performing a follow-up test (column 3). Finally, an adjustment to $\hat{x}$ is made in one fashion if the test result is positive and in another fashion if the result is negative (columns 4 and 5 respectively).

| Prevailing Condition | Corresponding Assumption | Follow-up Test | Positive Result | Negative Result |
|---|---|---|---|---|
| $\| \nabla f_{\mathbf{A}} u - \nabla \phi \| \geq \tau_{12}$ | $\hat{x} \in R_1$ | none | $\hat{x} \leftarrow \hat{x} + \alpha^* h^{\circ}$ | not applicable |
| $\leq \| \nabla f_{\mathbf{A}} u - \nabla \phi \| < \tau_{12}$ | $\hat{x} \in R_2$<br>and<br>no drop | $\| u_j \| \leq 1$, all $j \in \mathbf{A}$<br>or<br>$\| u_{j_0} \| > 1$, some $j_0 \in \mathbf{A}$<br>but<br>$d^T \nabla \phi \geq -\delta_1$ | $\hat{x} \leftarrow \hat{x} + \alpha^* h^{\circ}$<br>followed by<br>$\hat{x} \leftarrow \hat{x} + v$<br>if<br>$F(\hat{x} + v) < F(\hat{x}) - \delta_2 \| f_{\mathbf{A}} \|$ | $\hat{x} \leftarrow \hat{x} + \alpha^* d$ |
| $\nabla f_{\mathbf{A}} u - \nabla \phi \| < \tau_{23}$<br>and<br>sitive Result of table<br>lines 2 or 3<br>ocurred last step<br>and<br>$\hat{x} \leftarrow \hat{x} + v$<br>as successful in the<br>case of line 2 | $\hat{x} \in R_3$ | $F(\hat{x} + h^{\circ\circ} + v)$<br>$< F(\hat{x}) - \delta_3$ | $\hat{x} \leftarrow \hat{x} + h^{\circ\circ} + v$ | $\hat{x} \leftarrow \hat{x} + \alpha^* h^{\circ\circ}$<br>followed by<br>$\hat{x} \leftarrow \hat{x} + v$<br>if<br>$F(\hat{x} + v) <$<br>$F(\hat{x}) - \delta_2 \| f_{\mathbf{A}} \|$ |

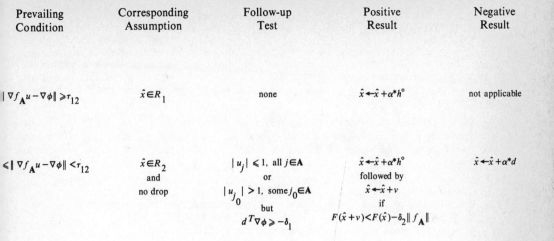

In addition to the above, $\epsilon, \tau_{12}, \tau_{23}$ are reduced whenever $F(\hat{x}+v)<F(\hat{x})-\delta_2\| f_{\mathbf{A}}\|$ or $F(\hat{x}+h^{\infty}+v)<F(\hat{x})-\delta_3$ fail to hold. Whenever linear dependencies in $\nabla f_{\mathbf{A}}$ are detected while computing $d$ or whenever the condition $|f_{j_0}(\hat{x}+\alpha^* d)|<\epsilon$ is detected following the "Negative Result" outcome in line 2 then $\epsilon$ alone is reduced. The reduction of $\epsilon$ and the $\tau$'s is heuristic (division by two). The reduction of $\epsilon$ in the second case is just enough to shrink $\mathbf{A}$ so that the gradients $\nabla f_j\, j\in\mathbf{A}$ are independent, and in the third case just enough to ensure that $j_0$ will not reappear in $\mathbf{A}$ at the next pass through the table, i.e. just enough to ensure that $f_{j_0}$ is properly "dropped".

Let $k\in\{1,2,3,\ldots\}$ index the sequence of iterations, and let $\hat{x}^{(k)}$, $\epsilon^{(k)}$, $\mathbf{A}^{(k)}$, $\tau_{12}^{(k)}$ and $\tau_{23}^{(k)}$ denote the $\hat{x}$, $\epsilon$, $\tau$'s and $\mathbf{A}$ encountered. Under the 7 assumptions stated earlier, with the set $\mathbf{S}$ mentioned therein, and for all $\eta>0$ in (2.14) sufficiently small, the discussions in [3,4] lead to the following

**Results**

a. The sequences of values $\epsilon^{(k)}$, $\tau_{12}^{(k)}$ and $\tau_{23}^{(k)}$ are bounded away from zero, and the elements of the sequence of index sets $\mathbf{A}^{(k)}$ are identical for all $k$ sufficiently large.

b. There exists a first-order point $\bar{x}\in\mathbf{S}$ such that $\lim_{k\to\infty}\hat{x}^{(k)}=\bar{x}$.

c. For all $k$ sufficiently large, only the positive outcome of line 3 in the table occurs (i.e only Newton steps are taken).

d. $\lim_{k\to\infty}\dfrac{\|\hat{x}^{(k+1)}-\bar{x}\|}{\|\hat{x}^{(k-1)}-\bar{x}\|}=0$, implying superlinear convergence.

**4. Test Problems**

Some representative problems from [5,6] have been tested. They are referred to below by the labels they were given in these references. In each case we report the number of evaluations of the function $F$ and the number of iteration steps of the algorithm $(x\to x+\Delta x)$ which were required to go from the stated initial point to a point having 7-place agreement with the solution (single-precision accuracy on our computer, the Honeywell 6600). It should be noted that our method, exclusive of what the line search requires, needs one evauation of all $f_i$, $\nabla f_i$ (that is, one evaluation of $F$ and $\nabla\phi$) together with one updated version of the matrix $Q$. (approximating $\nabla^2\phi$ or $B$ according to the nearness of $x$ to a stationary point) for each iteration. In addition, if the iteration step involves a vertical segment $v$, a further evaluation of $f_j$, $j\in\mathbf{A}$, is required. Any disparity shown below between the number of function evaluations and the number of iteration steps is the responsibility of the line search used -- an exceedingly crude and simplistic one in our current code.

Where possible we have tried to give a comparative figure for function evaluations and iteration steps as they are indicated in [5,6]. It will be seen from this that our method appears to be comparable to the one described in [6] and significantly better than the one described in [5]. Further computational testing and refinement are in progress.

Problem 1 ([5] Example 5.1):

$f_1(x)=x_1^2+x_2-10$

$f_2(x)=x_1+x_2^2-7$

$f_3(x)=x_1^2-x_2^3-1$

Initial Point: $x_1=1$, $x_2=2$

Minimal Point:

$x_1=2.842503$,

$x_2=1.920175$

Minimum $F$ Value: 4.704243E-01

Function Evaluations: 14

Iteration Steps: 6
(The algorithm of [5] had not attained comparable accuracy after 12 iterations and 88 function evaluations.)

Problem 2 ([5] Example 5.2, [6] Problem 4):

$$f_1(x) = x_1^2 + x_2^2 = x_3^2 - 1$$

$$f_2(x) = x_1^2 + x_2^2 + (x_3 - 2)^2$$

$$f_3(x) = x_1 + x_2 + x_3 - 1$$

$$f_4(x) = x_1 + x_2 - x_3 + 1$$

$$f_5(x) = 2x_1^3 + 6x_2^2 + 2(5x_3 - x_1 + 1)^2$$

$$f_6(x) = x_1^2 - 9x_3$$

Initial Point: $x_1 = 1,\ x_2 = 1,\ x_3 = 1$
Minimal Point:

$x_1 = 0.5360725$
$x_2 =$ -3.2E-10 $\approx 0$
$x_3 = 0.03193041$

Minimum $F$ Value: 7.894227
Function Evaluations: 20
Iteration Steps: 10
(The algorithm of [5] attained a comparable value of $F$ only after 11 iteration steps and 109 function evaluations. The results in [6] are recorded only to 5 figures. Agreement in the minimal $F$ value to that number of figures was reported attained after 10 iterations and 11 function evaluations.)

Problem 3 ([5] Example 6.1):

$$y = y(t) = \tfrac{1}{2}e^{-t} - e^{-2t} + \tfrac{1}{2}e^{-3t} + \tfrac{3}{2}e^{-3t/2}\sin(7t) + e^{-5t/2}\sin(5t)$$

$$h(t;x) = x_1 e^{-x_2 t}\cos(x_3 t + x_4) + x_5 e^{-x_6 t}$$

$$f_i(x) = h(t_i;x) - y(t_i) = h(t_i;x) - y_i$$

where $t_i = 0.0 + (i-1)/10.0,\ i = 1, \ldots, 51.$
Initial point:

$x_1 = 2,\ x_2 = 2,\ x_3 = 7$
$x_4 = 0,\ x_5 = -2,\ x_6 = 1$

Minimal Point:

$x_1 = -2.240744$
$x_2 = 1.857688$
$x_3 = 6.770049$
$x_4 = 1.496694$
$x_5 = 0.1658920$
$x_6 = 0.7422845$

Minimum $F$ Value: 0.5598131
Function Evaluations: 78
Iteration Steps: 68
(The Algorithm of [5] converges to a different minimal point having the same minimum $F$ value. After 11 iterations and 116 function evaluations it had attained the $F$ value 0.559818.)

Problem 4 ([6] Problem 3):

$f_1(x) = x_1^2 + x_2^2 + x_1 x_2$

$f_2(x) = \sin(x_1)$

$f_3(x) = \cos(x_2)$

Initial Point: $x_1 = 3, x_2 = 1$
Minimal Point:
$x_1 = 0.000000$
$x_2 = -8.3E\text{-}04 \approx 0.0$
Minimum $F$ Value: 1.000000
Function Evaluations: 20
Iteration Steps: 16
(The Algorithm of [6] reports termination at the point $x_1 = 0$, $x_2 =$ 2.0E-04 after 15 iterations and 15 function evaluations.)

## 5. Of Things Not Mentioned

One would expect $l_1$ minimization to be used most frequently in the context of data fitting; e.g. the simple linear model problem in the first section or test problem 3 above. Moreover, $l_1$ minimization will be chosen in these contexts over $l_2$ minimization often because there is a suspicion about the validity of a small portion of the data; e.g. the simple linear model problem. The vast majority of the data may be excellent, and this will mean that the vast majority of the functions $f_i(x) = h(t_i; x) - y_i$ may be nearly zero at some points $x$ which the algorithm must consider. The gradients $\nabla f_i$ are likely to be linearly dependent at such points; i.e. the algorithm will have to contend with degeneracy. We believe that horizontal, vertical and Newton steps remain reasonable and retain their theoretic properties under some weaker constraint qualification than that of linear independence (Assumption 1) if the choice of $\mathbf{A}$ is taken to be a subset of $\mathbf{A}(x, \epsilon)$ associated with a spanning collection of the vectors $\nabla f_j(x)$, $j \in \mathbf{A}(x, \epsilon)$. A similar situation is not evident for dropping steps. Thus, the algorithm as described could be expected to have dificulty in $R_2$ regions about degenerate stationary points. This is an area of further research.

## Bibliography


[1] Bartels, R.H. and Conn, A.R. Linearly constrained discrete $l_1$ problems. ACM Trans. on Math. Software 6 (1980) 594-608

[2] Coleman, T.F. and Conn, A.R. Second-order conditions for an exact penalty function. Math. Prog. 19 (1980) 178-185.

[3] Coleman, T.F. and Conn, A.R. Nonlinear programming via an exact penalty function: global analysis. Math. Prog. *to appear.*

[4] Colman, T.F. and Conn, A.R. Nonlinear programming via an exact penalty function: asymptotic analysis. Math. Prog. *to appear.*

[5] El-Attar, R.A.; Vidyasagar, M. and Dutta, S.R.K. An Algorithm for $l_1$-norm minimization with application to nonlinear $l_1$-approximation. SIAM J. Num. Anal. 16 (1979) 70-86.

[6] Overton, M.L. and Murray, W. A Projected Lagrangian algorithm for nonlinear $l_1$ optimization. SIAM J. Sci. and Stat. Comp. *to appear.*

# TOWARDS A UNIFIED APPROACH TO DATA SMOOTHING

J.L. FARAH
IIMAS - UNAM
Apdo. Postal 20-726
México 20, D.F.
MEXICO

## 1. INTRODUCTION

The basic principle underlying most of the literature on smoothing can be said to be based on the following formulation [21].

"In order to find a sounder basis of the theory, we must remember that the problem of graduation belongs essentially to the mathematical theory of probability;...the given data would constitute the 'most probable' values for the corresponding values of the argument, were it not that we have a priori grounds for believing that the true values form a smooth sequence, the irregularities being due to accidental causes which it is desirable to eliminate"

The mathematical machinery developed within the cited reference for the purpose of expressing this idea, is mainly this:

If our data are the numbers $\{y_t : t \in Z\}$ ($Z$: set of integers), it is desired to find numbers $\{\eta_t : t \in Z\}$ such that we would like to maximize the probability:

$$P = \exp\{-\frac{1}{2}\sum_t (y_t - \eta_t)^2\} \exp\{-\frac{1}{2\sigma^2}\sum_t [\Delta^3 \eta(t)]^2\}$$

(with $\Delta\eta(t) = \eta(t+1) - \eta(t)$)

Here, the maximization of $\exp\{-\frac{1}{2}\sum_t (y_t - \eta_t)^2\}$ would correspond to finding the "most likely" sequence $\{\eta\}$ yielding the sequence $\{y\}$ as our observations. The term $\sum_t [\Delta^3\eta(t)]^2$ has been traditionally used in actuarial practice to measure smoothness of a sequence $\{\eta\}$ (the smaller the quantity, the smoother $\{\eta\}$ is declared to be ). Moreover, while constructing P, the assumption is made that the sequence $\{\Delta^3\eta(t) : t \in Z\}$ of deviations from smoothness of our most likely sequence $\{\eta\}$, are to be regarded as sequence of independent gaussian irregularities with variance $\sigma^2$ and zero mean.

From here on, the solution is straightforward, in that the problem is reduced to minimizing the quadratic form:

$$J(\eta) = \frac{1}{2} \sum_t (y_t - \eta_t)^2 + \frac{1}{2\sigma^2} \sum_t [\Delta^3 \eta(t)]^2 \tag{1}$$

(The number 3 arises only for traditional purposes, as indicated above). More generally, one can consider a quadratic form: ($k \in \mathbb{N}$)

$$J(\eta) = \frac{1}{2} \sum_t (y_t - \eta_t)^2 + \lambda . \sum_t [\Delta^k \eta(t)]^2 ; \; \lambda > 0 \tag{2}$$

whose solution is equivalent to the solution of the doubly-infinite system of relations:

$$\eta_t + \lambda \cdot \Delta^{-k} \Delta^k \eta(t) = y_t \quad ; \quad t \in Z \tag{3}$$

(where $\Delta^{-1}\eta(i) = \eta(i) - \eta(i-1)$)

In what follows, it will be shown how this problem has appeared in the literature recurringly under different names and how it can be the subject of consideration of a variety of idiosyncrasies. (See Sect. 5).

## 2. A CLASS OF DIGITAL FILTERS

### a) Band-pass filters

In order to gain insight into the nature of the solution of (2) when regarded as a doubly infinite family of linear equations, ($i \in Z$), one reverts to the use of Laplace's method of generating functions [9] (z - transforms, [9 ]) as follows:

If $U(z) = \sum_{t \in Z} y_t z^{-t}$ converges within an annulus in the complex plane, $A = \{z \in C: \; r \leq |z| \leq R\}$, then $V(z) = \sum_{t \in Z} \eta_t z^{-t}$ should be given by the expression,

$$V(z) = \frac{1}{1 + \lambda (z^{-1} - 1)^k (z-1)^k} \; U(z) \tag{4}$$

which in engineering parlance [16], exhibits the relationship between input (U) and output (V) of a non-causal, linear, time-invariant "filter" whose transfer function is:

$$G_{\lambda,k}(z) = \frac{1}{1 + \lambda (z^{-1} - 1)^k (z-1)^k} \tag{5}$$

and whose "frequency characteristics" are given by the function $H_{\lambda,k}(\omega) = G_{\lambda,k}(e^{i\omega})$ ; $-\pi \leqslant \omega \leqslant \pi$.

From (5), it is immediate that:

(i) $H_{\lambda,k}(0)=1$ ; $H_{\lambda,k}(\pi)=(1+\lambda 2^{k+1})^{-1}$

(ii) $H_{\lambda,k}(\omega)< 0$; $0 < \omega < \pi$ ; $H^1_{\lambda,k}(0)=H^1_{\lambda,k}(\pi) = 0$

(iii) The higher k is, the higher is the number of zero derivatives of H at $\omega=0$. (The "flatter" H is at $\omega=0$).

(iv) If $\lambda > 2^{-k}$ ; $H_{\lambda,k}$ drops to 1/2 for $\omega$ equal to

$$\omega_{1/2}= \arc\cos(1- \frac{1}{2^k\sqrt{\lambda}})$$

(v) Given k, and any $\omega^*\epsilon(0,\pi)$, we can reach $H_{\lambda,k}(\omega^*)= \alpha$ $(0 < \alpha < 1)$ for a value of $\lambda(\lambda^*)$ computed by

$$\lambda^* = \frac{1-\alpha}{\alpha}[2(1-\cos\omega^*)]^{-k}$$

These five properties, show that we can interpret the smoothing process directly in terms of a band-pass filter with well-defined properties determined easily by the interplay of the two parameters k and $\lambda$.

b) Practical Implementation

In order to implement the filter in question, one proceeds as fol lows:

The polynomial $Q_\lambda(z)=1+\lambda(z^{-1}-1)^k(z-1)^k$ has no zeroes on the unit circle. Therefore, one can factorize it as

$$Q_\lambda(z) = R_\lambda(z)\, R_\lambda(z^{-1}) \tag{6}$$

with $R_\lambda(z)$ having all its zeroes within the unit circle. For this purpose one can use Bauer's algorithm [2] and thus obtain $R_\lambda(z)=\sum_{j=0}^{k} r_j z^j$. It is clear now that while

$$R_\lambda^{-1}(z) = \sum_{\ell=0}^{\infty} \rho_\ell z^{-\ell}; \quad |z| > 1 \tag{7}$$

we will have

$$Q_\lambda^{-1}(z) = \sum_{m=-\infty}^{+\infty} h_m z^{-m}; \text{ in an annulus A containing } \{|z|=1\} \tag{8}$$

From (7) it follows that $\rho_0=\rho_1=\ldots=\rho_{k-1}=0$ and that $\rho_k=r_k^{-1}$. From (8), it is seen that $h_{-m}=h_m$; $(m \geqslant 1)$ and that

$$h_m = \frac{1}{2\pi i} \int_{|z|=1} z^{m-1} \frac{d}{R_\lambda(z)\; R_\lambda(z^{-1})} \quad ; \; m \geqslant 0 \quad ; \tag{9}$$

and thus

$$\sum_{j=0}^{k} h_{m+j}\, r_j = \frac{1}{2\pi i} \int_{|z|=1} \frac{z^{m-1} dz}{R_\lambda(z^{-1})} \quad ; \; m \geqslant 0$$

and this integral is zero for all $m \geqslant 1$ (since $R_\lambda(z^{-1})$ has its zeroes outside the unit circle). (This technique is due to Doob [5 ]). From here, it follows that once we know $h_0, h_1, \ldots, h_k$, the values of $h_n$; $n \geqslant k+1$ are given recursively by:

$$r_k h_n + r_{k-1} h_{n-1} + \ldots + r_0 h_{n-k} = 0 \tag{10}$$

Combining (7) and (8) with (5) we obtain for $h_0, h_1, \ldots, h_k$ a linear system of the form:

$$\begin{array}{c} \\ k+1 \left\{ \begin{array}{c} \\ \\ \\ \\ \\ \end{array} \right. \\ \\ \\ k \left\{ \begin{array}{c} \\ \\ \\ \\ \\ \end{array} \right. \end{array} \left( \begin{array}{cccccccccc} r_k & r_{k-1} & \ldots & r_1 & r_o & 0 & \ldots & 0 & 0 \\ 0 & r_k & \ldots & r_2 & r_1 & r_o & \ldots & 0 & 0 \\ \cdot & \cdot & & \cdot & \cdot & \cdot & & \cdot & \cdot \\ \cdot & \cdot & & \cdot & \cdot & \cdot & & \cdot & \cdot \\ \cdot & \cdot & & \cdot & \cdot & \cdot & & \cdot & \cdot \\ 0 & 0 & \ldots & 0 & r_k & r_1 & \ldots & r_1 & r_o \\ \hline 1 & 0 & \ldots & 0 & 0 & 0 & \ldots & 0 & -1 \\ 0 & 1 & \ldots & 0 & 0 & 0 & \ldots & -1 & 0 \\ \cdot & \cdot & \ldots & \cdot & \cdot & \cdot & \ldots & \cdot & \cdot \\ \cdot & \cdot & \ldots & \cdot & \cdot & \cdot & \ldots & \cdot & \cdot \\ \cdot & \cdot & \ldots & \cdot & \cdot & \cdot & \ldots & \cdot & \cdot \\ 0 & 0 & \ldots & 1 & 0 & -1 & \ldots & 0 & 0 \end{array} \right)_{\underbrace{\qquad\qquad}_{2k+1}} \left( \begin{array}{c} x_1 \\ x_2 \\ \cdot \\ \cdot \\ \cdot \\ x_{k+1} \\ \hline x_{k+2} \\ x_{k+3} \\ \cdot \\ \cdot \\ \cdot \\ x_{2k+1} \end{array} \right) = \left( \begin{array}{c} \rho_o \\ \rho_1 \\ \cdot \\ \cdot \\ \cdot \\ \rho_k \\ \hline 0 \\ 0 \\ \cdot \\ \cdot \\ \cdot \\ 0 \end{array} \right) \tag{11}$$

and whose solution is such that $h_i = x_{k+1-i}$; $i=0,\ldots,k$.

The smoothing process will be now the convolution of $\{h\}$ with $\{y\}$. That is:

$$\eta_t = \sum_{m=-\infty}^{\infty} h_m\, y_{t-m} \tag{12}$$

which is performed in practice by truncating the summation (12) to a convenient value M of m, thus obtaining an approximating weighted moving average:

$$\eta_t = \sum_{m=-M}^{M} h_m y_{t-m} \tag{13}$$

c) Generalization

It is evident now that all the previous construction and analysis, (except for the very particular filter characteristics (i)-(v) in (b)) can be carried through in exactly the same way if, in (1) instead of $\Delta^k \eta(t)$, we write $S\eta(y)$ where S is a polynomial

$$S(\tau) = s_o + s_i \tau + \ldots + s_k \tau^k \tag{14}$$

and $\tau$ is the forward shift operator: $\tau\eta(t) = \tau(t+1)$ and we agree then to measure "smoothness" to be "large" if $\sum_t [S\eta(t)]^2$ is "small". To indicate how the analysis is carried out in this case, it suffices to mention that in this case, equations (2) are transformed into

$$\eta_i + \lambda\, S(\tau^{-1}) S(\tau) \eta(i) = y(i) \quad ; \quad i \in Z \tag{15}$$

where $\tau^{-1}\eta(t) = \tau(t-1)$.

The implementation in this case would again be carried out via an approximating weighted moving average of the form (13).

d) Truncation

Once formula (12) is considered it becomes obvious that for finite data, it is not simple to envisage how to apply it without loosing information at both ends of the data. From the point of view of Fourier analysis, if we regard a set of 2N+1 finite data $\{y_{-N}, \ldots, y_o, \ldots, y_N\}$ as the outcome of the infinite sequence $\{y_t : t \in Z\}$ "multiplied" with the sequence: $\{w_n : w_n = 1;\ -N \leqslant n \leqslant N;\quad w_n = 0;\ \text{ if } n < -N \text{ or } n > N\}$ then, the new transfer function, $G^{(N)}_{\lambda,k}$, is given by

$$G^{(N)}_{\lambda,k}(z) = \frac{1}{2\pi i} \int_{|\zeta|=1} G_{\lambda,k}(\zeta)\, D_N(z/\zeta) \frac{d\zeta}{\zeta} \tag{16}$$

where $D_N$ is the function giving rise to the Dirichlet kernel [16]:

$$D_N(w) = \begin{cases} \dfrac{w^N}{1-w^{-1}} + \dfrac{w^{-N}}{1-w} & ; \quad w \neq 1 \\ 2N+1 & ; \quad w = 1 . \end{cases}$$

Hence, the 'frequency response' is modified as:

$$G^{(N)}_{\lambda,k}(e^{i\omega}) = \frac{1}{2\pi} \int_0^{2\pi} G_{\lambda,k}(e^{i\alpha}) \frac{\operatorname{sen}\frac{1}{2}(2N+1)(\omega-\alpha)}{\operatorname{sen}\frac{1}{2}(\omega-\alpha)} d\alpha \qquad (17)$$

and, as a consequence, all the sharp frequency properties of $G_{\lambda,k}(e^{i\omega})$ are "smeared out" by the convolution (17). This effect is larger the smaller N is, and viceversa.

## 3. WEIGHTED MOVING AVERAGES

Because of the unpleasant effects given by (17) while processing finite data, Greville [ 7], poses the following problem; given data $\{y_p, y_{p+1}, \ldots, y_Q\}$ and a weighted moving average smoother (WMA):

$$\eta_t = \sum_{m=-M}^{M} h_m \, y_{t-m} \qquad (18)$$

can one find a way of conveniently extending the data {y} to some new values $y_{P-1}, y_{P-2}, \ldots, y_{P-M}$; $y_{Q+1}, y_{Q+2}, \ldots, y_{Q+M}$ so that a certain "consistency" is maintained between the chosen WMA and the data? His answer to this question is as follows:

a) A WMA is <u>exact for the degree r</u> if

(i) the sampled values $\{y_t = P(t);\ t \in Z\}$ of any polynomial P of degree r or less remain invariant after the application of the WMA.

(ii) r is the greatest integer with such a property.

As a consequence of this definition, r can only be of the form 2s-1; $s \leqslant M$ and the WMA has to have the operator form:

$$D(\tau) = 1 - \Delta^{-s}\Delta^{s} Q(\tau)$$

with

$$Q(\tau) = \sum_{j=-M+s}^{M-s} q_j \, \tau^j \quad ; \quad q_j = q_{-j} \qquad (19)$$

Further, the restriction is imposed on Q not to have zeroes on the unit circle.

b) Factorizing $Q(\tau)$ as $B(\tau)B(\tau^{-1})$ with B having all its M-s zeroes outside the unit circle (here again we use Bauer's algorithm [2 ]), and letting

$$A(\tau) = \Delta^s B(\tau) = \sum_{j=0}^{M} \alpha_j \tau^j \tag{20}$$

the WMA formula (18), becomes:

$$\eta_t = y_t - A(\tau^{-1})\, A(\tau) y(t) \tag{21}$$

c) Extension of Data. With (21), Greville proposes the following 'natural' conditions for computing extra data:

$$\begin{aligned} A(\tau)y_t &= 0 \quad ; \quad t = P-1,\dots,P-M \\ A(\tau^{-1})y_t &= 0 \quad ; \quad t = Q+1,\dots,Q+M \end{aligned} \tag{22a}$$

which yield $\eta_t = y_t$ if $t < p$ or $t > Q$. These conditions are reflected in the computation of $\eta_P, \eta_{P+1}, \dots, \eta_{P+M}$ and of $\eta_{Q-M+1}$, $\eta_{Q-M+2}, \dots, \eta_Q$ through the relations:

$$\begin{aligned} \eta_{P+i} &= y_{P+i} - \sum_{j=0}^{i} \alpha_j \; A(\tau) \quad y_{P+i-j} \\ \eta_{Q-i} &= y_{Q-i} - \sum_{j=0}^{i} \alpha_j \; A(\tau^{-1}) \; y_{Q-i+j} \end{aligned} \qquad i=0,1,\dots,M-1 \tag{22b}$$

and the smoothing algorithm is completed by directly applying (18):

$$\eta_t = \sum_{i=-M}^{M} h_j \, y_{t-i} = y_t - A(\tau^{-1})A(\tau)y(t) \tag{22c}$$

for $P+M \leqslant t \leqslant Q-M$. (Note that there is no actual need for performing the calculations in (22a)).

d) Matrix formulation. When expressed in matrix form, the relationships (22b) and (22c) become:

$$\eta = (I-F)y \quad ; \quad y \in \mathbb{R}^{N+1} \quad ; \quad \eta \in \mathbb{R}^{N+1} \tag{23}$$

where N=Q-P and F is the 'Trench matrix' ([7] [8]) whose elements $\varphi_{ij}$ are obtained by comparing coefficients in the following scheme: If

$$f_i(z) = \sum_{j=0}^{N} \varphi_{ij} z^j \quad ; \quad z \in C$$

then

$$f_i(z) = \begin{cases} z^i A(z) \sum_{j=0}^{i} \alpha_i z^{-j} & ; \quad 0 \leqslant i \leqslant M-1 \\ z^i A(z) A(z^{-1}) & ; \quad M \leqslant i \leqslant N-M \\ z^i A(z) \sum_{j=0}^{N-i} \alpha_j z^j & ; \quad N-M+1 \leqslant i \leqslant N \end{cases} \tag{24}$$

The structure (23) with (24) is important since an invertible Trench matrix has a banded Toeplitz matrix as its inverse [8], and thus formula (23) bears a strong tie ([7]) with the algorithms developed in the next section

## 4. SMOOTHING DIRECTLY IN $\mathbb{R}^N$

a) Problem Formulation

(i) Assume the data to be graduated, $\{y_i \leftarrow \mathbb{R}^N : i=1,\dots,N\}$ are considered as a point $y \in \mathbb{R}^N$ with coordinates $(y_1,\dots,y_N)$.

(ii) A k-dimensional linear subspace M of $\mathbb{R}^N$ is chosen a priori such that its elements are to represent smooth elements in $\mathbb{R}^N$. (Such an M, in fact a very simple one, can be constructed by considering the span of the vector all whose coordinates are equal to 1).

(iii) A linear map $S:\mathbb{R}^N \rightarrow \mathbb{R}^N$ is constructed such that its kernel is precisely M.

(iv) Two positive numbers $\mu,\lambda$ are a priori chosen so that the smoothing problem is formulated as the problem of finding $\eta^\circ \in \mathbb{R}^N$ which minimizes the function:

$$J(\eta) = \mu\|y-\eta\|_2^2 + \lambda\|S\eta\|_2^2 \tag{25}$$

such an $\eta^\circ$ exists uniquely and is given as the solution to the linear problem

$$[\mu I + \lambda S'S]\eta^\circ = \mu y \tag{26}$$

b) Invariance.

The solution to (26) can be written as

$$\eta^\circ = R_{\lambda,\mu}\ y \tag{27}$$

with $R_{\lambda,\mu} = (I + \frac{\lambda}{\mu}\ S'S)^{-1}$. Now we have the simple

Proposition

(i) 1 is an eigenvalue of $R_{\lambda,\mu}$ of (geometric) multiplicity k. Its associated eigenspace is $M(=\mathrm{Ker}(S))$.

(ii) Every other eigenvalue $\alpha$ satisfies the bounds:

$$1 - \frac{\lambda}{\mu}\,\|S\|_2^2 \leqslant \alpha < 1 \tag{28}$$

Proof

The proof is very easy in that, if $\alpha$ is an eigenvalue of $R_{\lambda,\mu}$ with v a correspoding eigenvector, we can write the identity:

$$\alpha(I + \frac{\lambda}{\mu}\ S'S)v = v \tag{29}$$

(i) if $v \in \ker(S)$ then $\alpha=1$ and conversely, if $\alpha=1$ we must have $S'Sv=0 \Rightarrow\ \in \mathrm{Ker}(S')=[R(S)]^{\perp}=M$.

(ii) If $v \notin \mathrm{Ker}(S)$ then, from (29)

$$\alpha = 1 - \frac{\lambda}{\mu}\ \frac{\|Sv\|^2}{\|v\|^2} \tag{30}$$

and conversely, thus (28) follows.

Corollary: The mapping $R_{\lambda,\mu}$ contracts the complement of M if $\lambda$, $\mu$ are chosen so that

$$\frac{\lambda}{\mu} < 2\ \ \|S\|_2^{-2} \tag{31}$$

Remark

These results are the algebraic statements analogous to those of digital filters: (The elements of M remain intact by the smoothing process. The rest are reduced in norm)

c) Special Structures of S and Special Algorithms.

It is always possible to choose S so that its representing matrix (redundantly, we call it S as well) is of dimensions (N-k) x N:

$$S = \begin{bmatrix} \sigma_{11} & \sigma_{12} & \cdot\ \cdot & \sigma_{1N} \\ \sigma_{21} & \sigma_{22} & \cdot\ \cdot & \sigma_{2N} \\ \cdot & \cdot & \cdot\ \cdot & \cdot \\ \cdot & \cdot & \cdot\ \cdot & \cdot \\ \cdot & \cdot & \cdot\ \cdot & \cdot \\ \sigma_{N-k,1} & \sigma_{N-k,2} & \cdot\ \cdot & \sigma_{N-k,N} \end{bmatrix}$$

Of particular practical interest are matrices with an imbedded triangular Toeplitz structure:

$$S = \begin{bmatrix} \sigma_0 & \sigma_1 & \sigma_2 & \cdots & \sigma_{N-k-1} & \sigma_{N-k} & \cdot & \sigma_{N-1} \\ 0 & \sigma_0 & \sigma_1 & \cdots & \sigma_{N-k-2} & \sigma_{N-k-1} & \cdots & \sigma_{N-2} \\ \cdot & \cdot & \cdot & & \cdot & \cdot & & \cdot \\ \cdot & \cdot & \cdot & & \cdot & \cdot & & \cdot \\ \cdot & \cdot & \cdot & & \cdot & \cdot & & \cdot \\ 0 & 0 & 0 & \cdots & \sigma_0 & \sigma_1 & \cdots & \sigma_k \end{bmatrix}$$

with the properties:

(i) The left (N-k)x(N-k) (Toeplitz) block of S commutes with the power of the "shift-operator" H whose elements are all zero except for the first upper sub-diagonal whose elements are all 1. This is easily seen if one regards the block in question as the matrix q(H), where q is the polynomial

$$q(t) = \sigma_0 + \sigma_1 t + \ldots + \sigma_{N-k}\, t^{N-k} \tag{33}$$

This gives S a certain "shift-invariant" property.

(ii) The rows of (32) form a basis for the annihilator subspace $M^\circ$ of M if rank(S)=N-k.
If in particular, q(t) in (33) is chosen as $(t-1)^k$, then the kernel of S is spanned by the values of polynomials of degree $\leqslant$ k-1 evaluated on an equidistant mesh [cf. section 3], and one obtains the matrix: $(N \geqslant k)$

$$S_N = \begin{bmatrix} \sigma_0 & \sigma_1 & \cdots & \sigma_k & 0 & \cdots & 0 & 0 & \cdots & 0 \\ 0 & \sigma_0 & \cdots & \sigma_{k-1} & \sigma_k & \cdots & 0 & 0 & \cdots & 0 \\ \vdots & \vdots & & \vdots & & & \cdot & \vdots & & \vdots \\ 0 & 0 & & 0 & 0 & \cdots & \sigma_0 & \sigma_1 & \cdots & \sigma_k \end{bmatrix} \tag{34}$$

In this case, $S_N'S_N$ is a Trench matrix: (23)

This last structure, allows for the design of recursive (memory-saving) algorithms as follows:

If one defines, (following eq. (26)) $U_N=\mu I_N+ S_N'S_N$, then, introducing the notions: (for $N > k$)

$$\eta_N = U_N^{-1} \cdot y_N \ ; \ \pi_N' = (0,\dots,0 \quad \sigma_0 \dots \sigma_{k-1})$$
$$z_N = U_N^{-1} \cdot \pi_N \ ; \ \tilde{\pi}_N' = (\sigma_1,\dots,\sigma_k \quad 0\dots 0 \quad ) \tag{35}$$
$$\tilde{z}_N = U_N^{-1} \cdot \tilde{\pi}_N'$$

(all of these are vectors in $\mathbb{R}^N$) ; one ends up with

$$U_{N+1} = \left[\begin{array}{c|c} U_N + \lambda\pi_N\pi_N' & \lambda\sigma_k\pi_N \\ \hline \lambda\sigma_k\pi_N' & \mu+\lambda\sigma_k^2 \end{array}\right] = \left[\begin{array}{c|c} \mu + \lambda\sigma_0^2 & \lambda\sigma_0\tilde{\pi}_N' \\ \hline \lambda\sigma_0\tilde{\pi}_N & U_N+\lambda\tilde{\pi}_N\tilde{\pi}_N' \end{array}\right] \tag{36}$$

and, if further, one defines the scalars:

$$\rho_N = \pi_N'U_N^{-1}\pi_N = \pi_N'z_N \quad ; \quad \tilde{\rho}_N = \tilde{\pi}_N'U_N^{-1}\tilde{\pi}_N = \tilde{\pi}_N'\tilde{z}_N \tag{37}$$

the following recursions are gotten:

$$\eta_{N+1} = \left[\begin{array}{c} \eta_N - \dfrac{\mu\rho_N-\sigma_k y^{(N+1)}}{\lambda\sigma_k^2+\mu(1+\lambda\rho_N)} \cdot z_N \\ \hline \dfrac{(1+\lambda\rho_N)y^{(N+1)}-\lambda\sigma_k\rho_N}{\lambda\sigma_k^2+\mu(1+\lambda\rho_N)} \end{array}\right] \tag{38a}$$

$$z_{N+1} = \left[\begin{array}{c} -\dfrac{\lambda\sigma_0}{\lambda\sigma_0^2+\mu(1+\lambda\tilde{\rho}_N)} \tilde{\pi}_N'z_N \\ \hline z_N-\dfrac{\lambda\mu}{\lambda\sigma_0^2+\mu(1+\lambda\tilde{\rho}_N)} (\tilde{\pi}_N'z_N)\tilde{z}_N \end{array}\right] \tag{38b}$$

$$\tilde{z}_{N+1} = \left[\begin{array}{c} \tilde{z}_N-\dfrac{\lambda\mu}{\lambda\sigma_0^2+\mu(1+\lambda\rho_N)} (\pi_N'\tilde{z}_N)z_N \\ \hline -\dfrac{\lambda\sigma_k}{\lambda\sigma_0^2+\mu(1+\lambda\rho_N)} (\pi_N'\tilde{z}_N) \end{array}\right] \tag{38c}$$

($y^{(N+1)}$ is the $(N+1)^{st}$ coordinate of $y_{N+1}$), where one starts with the initial quantities:

$$S_{k+1} = s' = (\sigma_0, \sigma_1, \ldots, \sigma_k)$$

$$\pi'_{k+1} = (0, \sigma_0, \ldots, \sigma_{k-1}) \quad ; \quad \tilde{\pi}'_{k+1} = (\sigma_1, \ldots, \sigma_k, 0) \tag{38d}$$

and

$$U^{-1}_{k+1} = \frac{1}{\mu} I_{k+1} - \frac{\lambda}{\mu + \lambda s's} ss'$$

The recursive scheme (38 a-d) is obtained by repeatedly manipulating rank-one corrections to matrix inverses and by exploiting the partitions (36) of $U_{N+1}$. (Thus an algorithm very much like the algorithm of Trench [20] for inverting Toeplitz matrices is obtained).

## 5. CONCLUDING REMARKS

The modest theory expounded in section 4 lies at the very heart of several developments which have appeared in the literature on the subject, and all these approaches differ only in the way that eqn. (24) is considered:

a) If eqn. (24) is properly posed in Hilbert spaces, one obtains the theory of spline-smoothing, [17], [18], [11], and its relation to filtering has also been explored [4].

b) If the triangular operator q(H) is regarded (in function space) as giving rise to a dynamical system, a link can be established between control problems and smoothing [1], [3], [19]. The idea can be traced back to Pontryagin [15].

c) As in (b), when one considers a probabilistic argument and a dynamical system structure, one ends up with the theory of smoothing à la Kalman [10], [13], [12], [19], [14]. It is well known that between Kalman Filtering and Control Theory there exists a certain duality which allows one to pass from a Filtering Problem (in this case, smoothing) to a Control Problem and viceversa.

## REFERENCES

[1] Aronsson, G., "Perfect Splines and Nonlinear Optimal Control Theory", J. of Approx. Th., 25, 142-152 (1979).

[2] Bauer, F.L., "Beiträge zur Entwicklung numerischer Verfahren für programmgesteuerte Rechenanlagen: II. Direkte Faktorisierung eines Polynoms", Bayerische Akad. d. Wiss., München (1956).

[3] Bellman, R., Kashef, B.F., and Vasudevan, R., Splines via Dynamic Programming. JMAA 38, 471-479 (1972).

[4] De Figuereido, R.J.P., "LM-g splines", Technical Report No. 7510. Dept. of EE, Rice University, Houston, Tex. (1975).

[5] Doob, J.L., "Stochastic Processes", Wiley, (1953). Page 503 is the one referred to in the text .

[6] Duris, C.S., "Discrete Interpolating and Smoothing Spline Functions", SIAM J. Numer. Anal., 14, 4, (1977).

[7] Greville, T.N.E., "Moving-Weighted-Average Smoothing Extended to the Extremities of the Data", MRC Tech. Summary Rep. No. 2025, Nov. (1979).

[8] Greville, T.N.E., Trench, W.F., "Band Matrices with Toeplitz Inverses", Lin. Alg. & its Appl., 27, 199-209 (1979).

[9] Jury, E.I., "Theory and Application of the z-transform Method", J. Wiley, N.Y. (1964).

[10] Kalman, R.E., "New Methods and Resutls in Linear Prediction and Filtering Theory", Research Inst. of Adv. Studies, Report No. 61-1, (1961).

[11] Laurent, P.J., "Approximation et Optimisation", Hermann, (1972).

[12] Lindquist, A., "A Theorem on Duality Between Estimation and Control for Linear Stochastic Systems with Time Delay", J. Math. Anal. Appl., 37, 516-536 (1972).

[13] Mayne, D.Q., "A Solution of the Smoothing Problem for Linear Dynamic Systems", Automatica, 4, 73-92 (1966).

[14] Badawi, F.A., Lindquist, A., Pavon, M., "A Stochastic Realization Approach to the Smoothing Problem", IEEE Trans. AC, AC-24, 6, 878-888 (Dec. 1979).

[15] Pontryagin, L.S., "Theorie Mathématique des Processus Optimaux", Trad, Francaise, Mir., (1974). [Russian Edition, 1959].

[16] Rabiner, L.R., "Theory & Application of Digital Signal Processing". Prentice-Hall (1975).

[17] Reinsch, C.H., "Smoothing by Spline Functions", Numer. Math., 10, 177-183 (1967).

[18] Reinsch, C.H., "Smoothing by Spline Functions II", Numer. Math., 16, 451-454 (1971).

[19] Sidhu, G.S., Desai, U.B., New Smoothing Algorithms Based on Reversed-Time Lumped Models", IEEE, Trans. on AC., AC-21, 538-541, (1976).

[20] Trench, W.F., "An Algorithm for the Inversion of Finite Toeplitz Matrices", J. SIAM, 12, 515-422 (1964).

[21] Whittaker, E., Robinson, G., "The Calculus of Observations", (an Introduction to Numerical Analysis), (1924), 4th. Edition, Dover, (1967).

# GENERALIZING THE LINPACK CONDITION ESTIMATOR

Alan K. Cline
Department of Computer Science
University of Texas
Austin, Texas 78712

Andrew R. Conn
Department of Computer Science
University of Waterloo
Waterloo, Ontario N2L 3G1

Charles F. Van Loan
Department of Computer Science
Cornell University
Ithaca, New York 14853

## 1. Background

Suppose the n-by-n nonsingular linear system $Ax = b$ is solved using a "stable" matrix factorization method such as Gaussian elimination with pivoting. If t-digit, base $b$ floating point arithmetic is used then it is generally the case that the relative error in the computed solution $\hat{x}$ satisfies

$$\frac{\| \hat{x} - x \|_p}{\| x \|_p} \cong b^{-t} k_p(A) . \tag{1}$$

Here, $\| x \|_p = (|x_1|^p + \ldots + |x_n|^p)^{1/p}$ and $k_p(A)$ is the p-norm condition of A defined by

$$k_p(A) = \max_{z \neq 0} \frac{\| Az \|_p}{\| z \|_p} \Big/ \min_{z \neq 0} \frac{\| Az \|_p}{\| z \|_p} \equiv \| A \|_p \, \| A^{-1} \|_p .$$

Heuristic (1) implies that an estimate $\hat{k}_p$ to $k_p(A)$ can be useful when assessing the quality of $\hat{x}$ .

Among other things, the attractiveness of a condition estimator depends upon its reliability and how expensive it is to compute. With respect to reliability, we adopt the convention that $\hat{k}_p$ is a reliable estimator if

$$c_1 k_p(A) \leq \hat{k}_p \leq c_2 k_p(A) \tag{2}$$

for "reasonable" constants $c_1$ and $c_2$ independent of A .

Consider, for example, the estimator $\hat{k}_2 = \hat{\sigma}_1 / \hat{\sigma}_n$ where $\hat{\sigma}_1 \geq \ldots \geq \hat{\sigma}_n \geq 0$ are the singular values of A computed by either the EISPACK [7] or LINPACK [2] singular value decomposition (SVD) subroutine. This estimate is provably reliable in that it can be rigorously shown that the constants in (2) are each of the form $1 + O(b^{-t})$. Unfortunately, computing the SVD requires about 15 times as many flops as Gaussian elimination and so this is a rather expensive method for assessing $\hat{x}$. (A "flop" is a floating point multiplicative operation.) Condition estimators that require only $O(n^2)$ flops once we have computed a "cheap" factorization such as PA = LU (via Gaussian elimination with partial pivoting) or AP = QR (via Householder triangularization with column pivoting) are therefore of interest.

Note that with such a factorization available we can readily compute the estimate $\|A\|_p \|\hat{X}\|_p \approx k_p(A)$ where $p = 1$ or $\infty$ and $\hat{X}$ is the computed solution to AX = I. Although estimators of this type are provably reliable, they are inefficient because they require $O(n^3)$ flops. The challenge, therefore, is to reliably estimate the condition in $O(n^2)$ flops assuming that A has already been factored.

Forsythe and Moler [3] propose an interesting $O(n^2)$ estimator based on iterative improvement and the assumption that $\hat{x}$ has been computed via PA = LU. In particular, they set $\hat{k}_\infty = b^t \|\hat{z}\|_\infty / \|\hat{x}\|_\infty$ where $\hat{z}$ is computed by solving Lw = Pr and Uz = w and where the residual $r = b - A\hat{x}$ is calculated in double precision. Although this estimator is efficient, its attractiveness is limited (a) because of portability problems associated with the double precision calculation of r and (b) because it is necessary to have an extra n-by-n array. Its reliability is unproven but it appears to be a very succesful technique.

Karasalo [5] describes an efficient 2-norm condition estimator that is based on the properties of the upper triangular matrix R that is computed via Householder triangularization with column pivoting. However, the estimator is not provably reliable to the extent that the constant $c_1$ in (2) must be of order $4^{-n}$.

## 2. The LINPACK Approach

Another approach to the condition estimation problem is taken by Cline, Moler, Stewart, and Wilkinson [1] and is implemented in LINPACK. It amounts to inverse iteration with a special technique for choosing the starting vector. The method assumes that A has been factored and proceeds as follows:

Step 1. Choose d such that the solution to $A^T w = d$ is large in norm relative to d.

Step 2. Solve Az = w.

Step 3. Set $k_1 = \|A\|_1 \|z\|_1 / \|w\|_1$

Note that since $\|A^{-1}\|_1 \geq \|z\|_1 / \|w\|_1$ we have

$$\frac{\| z \|}{\| A^{-1}\|_1 \| w \|_1} k_1(A) = \hat{k}_1 \leq k_1(A)$$

Hence, from the standpoint of reliability, it is desirable that $\| z \|_1 / \| w \|_1$ be as close to $\| A^{-1}\|_1$ as possible. That Step 1 encourages this can be seen via a brief 2-norm argument using the SVD. Let

$$A = U \operatorname{diag}(\sigma_i) V^T \quad , \quad U^T U = I \ , \ V^T V = I \ , \quad \sigma_1 \geq \sigma_2 \geq \dots \geq \sigma_n \geq 0$$

be the SVD of $A$ with $U = [u_1, \dots, u_n]$ and $V = [v_1, \dots, v_n]$ . If

$$d = \sum_{i=1}^{n} a_i v_i$$

then

(3) $$w = \sum_{i=1}^{n} \frac{a_i}{\sigma_i} u_i \qquad \text{and} \qquad z = \sum_{i=1}^{n} \frac{a_i}{\sigma_i^2} v_i \ .$$

A calculation shows that

$$\frac{\| z \|_2^2}{\| w \|_2^2} \geq \frac{a_n^2}{\sigma_n^2 \| d \|_2^2} \cdot \frac{\sum_{i=1}^{n} a_i^2}{\sum_{i=1}^{n} \left[\frac{\sigma_n}{\sigma_i}\right]^2 a_i^2}$$

i.e.,

$$\frac{\| z \|_2}{\| w \|_2} \geq \| A^{-1} \|_2 \cos(v_n, d) \quad , \quad \cos(v_n, d) = \frac{|a_n|}{\| d \|_2}$$

Thus, it is desirable that $\cos(v_n, d)$ be near unity. As (3) suggests, striving for a large $w$ in Step 1 tends to produce a vector $d$ that has a significant component in the direction of $v_n$ .

To motivate the LINPACK method for carrying out Step 1, assume that $T$ is an n-by-n lower triangular matrix and consider the problem of choosing $d$ such that the solution to $Ty = d$ has a large norm. Since $y$ can be computed as follows,

$p_k := 0 \quad (k = 1, \dots, n)$

For $k = 1, \dots, n$

$\quad y_k := (d_k - p_k)/t_{kk}$

$\quad p_i := p_i + t_{ik} y_k \quad (i = k+1, \dots, n)$

it is clearly desirable that $d_k$ be chosen such that both $y_k$ and the running sums $p_{k+1},\ldots,p_n$ are as large as possible. This can be done by setting $d_k = a$ where $a \in \{-1,+1\}$ maximizes

$$\phi_k(a) = |y_k(a)| + \sum_{i=k+1}^{n} w_i \, |p_i + t_{ik}\, y_k(a)|$$

Here, $y_k(a) = (a - p_k)/t_{kk}$ and the $w_i$ are nonnegative weights. In LINPACK, the weights are all set to one. Another option mentioned in [1] is to set $w_i = 1/t_{ii}$.

If $A$ is square and $PA = LU$, then the LINPACK estimator determines the vector $d$ in Step 1 by applying the above scheme with $T = U^T$. Note that Steps 1-3 require $O(n^2)$ flops so the method is efficient. However, its success depends on an additional heuristic. Namely, that by striving for a large norm solution to $U^T y = d$ we obtain a large norm solution to $A^T w = d$. Experimental evidence suggests that ill-conditioning in $A$ tends to be reflected in ill-conditioning in $U$ and so there is some justification for this approach. We will comment more fully on the method's reliability in Section 6.

## 3. Estimators with "Look Behind"

In this and the next section we assume that $A = T$ is lower triangular and we consider various alternatives to the LINPACK method for producing a large norm solution to $Ty = d$. Our first alternative incorporates the notion of "look behind" and we begin by developing a 2-norm condition estimator that has this feature. For the sake of clarity, assume that $n=6$ and that $d_1, d_2$, and $d_3$ are known and satisfy $d_1^2 + d_2^2 + d_3^2 = 1$. Also assume that that we have solved the system

$$\begin{bmatrix} t_{11} & 0 & 0 \\ t_{21} & t_{22} & 0 \\ t_{31} & t_{32} & t_{33} \end{bmatrix} \begin{bmatrix} y_1 \\ y_2 \\ y_3 \end{bmatrix} = \begin{bmatrix} d_1 \\ d_2 \\ d_3 \end{bmatrix} \qquad (4)$$

and have computed the "look ahead" values

$$\begin{aligned} p_4 &= t_{41}y_1 + t_{42}y_2 + t_{43}y_3 \\ p_5 &= t_{51}y_1 + t_{52}y_2 + t_{53}y_3 \\ p_6 &= t_{61}y_1 + t_{62}y_2 + t_{63}y_3 \end{aligned}$$

We now determine $c = \cos(a)$ and $s = \sin(a)$ such that if

$$\begin{bmatrix} t_{11} & 0 & 0 & 0 \\ t_{21} & t_{22} & 0 & 0 \\ t_{31} & t_{32} & t_{33} & 0 \\ t_{41} & t_{42} & t_{43} & t_{44} \end{bmatrix} \begin{bmatrix} y_1' \\ y_2' \\ y_3' \\ y_4' \end{bmatrix} = \begin{bmatrix} sd_1 \\ sd_2 \\ sd_3 \\ c \end{bmatrix}$$

then $\sum_{i=1}^{4} (y_i')^2 + \sum_{i=5}^{6} (p_i')^2$ is maximized where $p_i' = sp_i + t_{i4}y_4'$ , i=5,6.

(The $p_i'$ are updates of the $p_i$ .) Notice that by changing the righthand side in this fashion that the solution of the enlarged system is easily obtained:

$$y_4' = (c - sp_4)/t_{44}$$

$$y_i' = s\,y_i \qquad (i = 1,2,3)$$

Also observe that the new right hand side has unit 2-norm.

In general, at the k-th step when $d_k$ is computed, we "look behind" and consider the revision of $d_1,\ldots,d_{k-1}$ and we "look ahead" to anticipate the effects on $p_{k+1},\ldots,p_n$ . Overall we have

Algorithm 1

$p_k := 0 \quad (k = 1,\ldots,n)$

For k = 1,...,n

1. Determine $a \in [0,2\pi]$ such that if $c = \cos(a)$, $s = \sin(a)$, and $y_k(a) = (c - sp_k)/t_{kk}$ then

$$\phi_k(a) = s^2 \sum_{i=1}^{k-1} (y_i)^2 + y_k(a)^2 + \sum_{i=k+1}^{n} w_i^2[sp_i + t_{ik}y_k(a)]^2$$

is maximized where $w_1,\ldots,w_n$ are nonnegative weights.

2. $c := \cos(a)$; $s := \sin(a)$; $d_k := c$; $y_k := y_k(a)$

$d_i := s\,d_i \qquad (i=1,\ldots,k-1)$

$y_i := s\,y_i \qquad (i=1,\ldots,k-1)$

$p_i := s\,p_i + t_{ik}y_k \qquad (i=k+1,\ldots,n)$

The parameter a is easily determined. From the equation $\phi_k'(a) = 0$ we obtain the relation

(6) $$\beta c s = \alpha (c^2 - s^2)$$

where

$$\beta = (y^T y + p^T D^2 p) t_{kk}^2 + (p_k^2 - 1)(1 + t^T D^2 t) - 2 p_k t_{kk} p^T D^2 p$$

$$\alpha = p_k (1 + t^T D^2 t) - t_{kk} p^T D^2 t$$

and

$$t^T = (t_{k+1,k}, \ldots, t_{nk})$$

$$p^T = (p_{k+1}, \ldots, p_n)$$

$$y^T = (y_1, \ldots, y_{k-1})$$

$$D = \text{diag}(w_{k+1}, \ldots, w_n) .$$

The two possible sine-cosine pairs that satisfy (6) can be calculated as follows:

$$r := \beta / (2 \alpha)$$

$$\mu_1 := r + \text{sqrt}(1 + r^2) \; ; \quad s_1 := 1/\text{sqrt}(1 + \mu_1^2) \; ; \; c_1 := s_1 \mu_1 \; ;$$

$$\mu_2 := r - \text{sqrt}(1 + r^2) \; ; \quad s_2 := 1/\text{sqrt}(1 + \mu_2^2) \; ; \; c_2 := s_2 \mu_2 \; ;$$

Which pair maximizes can be deduced upon substitution into $\phi_k(a)$.

Algorithm 1 requires approximately $5n^2$ flops. It is readily seen to produce the estimate

(7) $$(\sigma_n, v_n, u_n) \cong (\hat{\sigma}_n, \hat{v}_n, \hat{u}_n) = ( \| y \|_2^{-1} , d , y / \| y \|_2 )$$

where $\sigma_n$ is the n-th singular value value of $T$ and $v_n$ and $u_n$ are the associated right and left singular vectors. On the other hand, if $a$ is chosen at each stage so as to minimize $\phi_k(a)$, then an estimate of the largest singular value and its singular vectors results:

(8) $$(\sigma_1, v_1, u_1) \cong (\hat{\sigma}_1, \hat{v}_1, \hat{u}_1) = ( \| y \|_2^{-1} , d , y / \| y \|_2)$$

Of course, (7) and (8) combine to give $\hat{k}_2 = \hat{\sigma}_1 / \hat{\sigma}_n$

An $L_1$ "look behind" condition estimator can also be devised. To illustrate, suppose $n=6$, $k=3$, and that equations (4) and (5) hold with $|d_1| + |d_2| + |d_3| = 1$. We seek $\lambda \in [0,1]$ such that if

$$\begin{bmatrix} t_{11} & 0 & 0 & 0 \\ t_{21} & t_{22} & 0 & 0 \\ t_{31} & t_{32} & t_{33} & 0 \\ t_{41} & t_{42} & t_{43} & t_{44} \end{bmatrix} \begin{bmatrix} y'_1 \\ y'_2 \\ y'_3 \\ y'_4 \end{bmatrix} = \begin{bmatrix} \lambda d_1 \\ \lambda d_2 \\ \lambda d_3 \\ 1 - \lambda \end{bmatrix}$$

then $\sum_{i=1}^{4} |y'_i| + \sum_{i=5}^{6} |p'_i|$ is maximized where $p'_i = \lambda p_i + t_{i4} y'_4$ , i=5,6 . Since the function to be maximized is convex, it suffices merely to check its value at $\lambda = 0$ and $\lambda = 1$ . In general we have

<u>Algorithm 2</u>

$p_k := 0 \quad (k = 1,\dots,n)$

For $k = 1,\dots,n$

1. Determine $\lambda \in \{-1,+1\}$ such that if

$$y_k(\lambda) = [(1 - \lambda) - \lambda p_k] \,/\, t_{kk}$$

then

$$\phi_k(\lambda) = \lambda \sum_{i=1}^{k-1} |y_i| + |y_k(\lambda)| + \sum_{i=k+1}^{n} w_i |\lambda p_i + t_{ik} y_k(\lambda)|$$

is maximized where the $w_i$ are nonnegative weights.

2. $d_k := 1 - \lambda$ ; $y_k := (d_k - \lambda p_k)/t_{kk}$

$d_i := \lambda d_i \quad (i=1,\dots,k-1)$

$y_i := \lambda y_i \quad (i=1,\dots,k-1)$

$p_i := \lambda p_i + t_{ik} y_k \quad (i=k+1,\dots,n)$

With $y$ computed in this fashion we obtain the estimate $\hat{k}_1 = \| T \|_1 \| y \|_1$ .

Note that the final right hand side $d$ is a column of the identity matrix and therefore, $y$ is some column of $T^{-1}$ . We remark that Algorithm 2 is considerably more efficient than Algorithm 1, especially since the parameter $\lambda$ is either zero or one.

## 4. A Divide and Conquer Estimator

Suppose $T_{11} \in R^{p \times p}$ and $T_{22} \in R^{q \times q}$ are lower triangular and that we have solved the systems

$$T_{11} y_1 = d_1 \qquad \| d_1 \|_2 = 1$$

$$T_{22} y_2 = d_2 \qquad \| d_2 \|_2 = 1$$

Consider the problem of choosing $c = \cos(a)$ and $s = \sin(a)$ such that if

$$\begin{bmatrix} T_{11} & 0 \\ T_{21} & T_{22} \end{bmatrix} \begin{bmatrix} z_1 \\ z_2 \end{bmatrix} = \begin{bmatrix} cd_1 \\ sd_2 \end{bmatrix}$$

then

$$\phi(a) = \| z_1 \|_2^2 + \| z_2 \|_2^2$$

is maximized. Define $w \in R^q$ by $T_{22}w = T_{21}y_1$ . A calculation shows that $z_1 = c\,y_1$ and $z_2 = s\,y_2 - c\,w$ and thus

(9) $$\phi(a) = c^2[\| y_1 \|_2^2 + \| w \|_2^2] - 2sc\, y_2^T w + s^2 \| y_2 \|_2^2$$

By manipulating the equation $\phi'(a) = 0$, we obtain the following method for determining the two sine-cosine pairs that give extreme values for $\phi(a)$ :

$$\beta := \| y_2 \|_2^2 - \| y_1 \|_2^2 - \| w \|_2^2$$

$$\alpha := y_2^T w$$

$$r := \beta / (2\alpha)$$

$$\mu_1 := r + \mathrm{sqrt}(1 + r^2) \; ; \quad s_1 := 1/\mathrm{sqrt}(1 + \mu_1^2) \; ; \quad c_1 := \mu_1 s_1 \; ;$$

$$\mu_2 := r - \mathrm{sqrt}(1 + r^2) \; ; \quad s_2 := 1/\mathrm{sqrt}(1 + \mu_2^2) \; ; \quad c_2 := \mu_2 s_2 \; ;$$

This computation forms the heart of a divide and conquer algorithm that can be used to produce a large norm solution to $Ty = d$ . Consider the case $n = 8$. We begin by solving the eight 1-by-1 systems $(t_{ii})y_i = 1$ . These systems are then paired and combined in the above fashion to produce four 2-by-2 systems that involve the matrices

$$\begin{bmatrix} t_{ii} & 0 \\ t_{ji} & t_{jj} \end{bmatrix} \qquad i = 1,3,5,7 \; ; \; j = i+1$$

These systems are in turn paired and combined, all the while choosing the sines and cosines to encourage growth. Finally, the two 4-by-4 systems are synthesized to render a final $y$ and $d$ satisfying $Ty = d$.

In the general case there are several ways to handle the pairing of the systems in the event that $n$ is not an exact power of 2. Our approach is as follows. Suppose at some stage we have $k$ linear systems $S_1,\ldots,S_k$ and that these are to be paired together. Write $k = 2p + q$ where $q$ is either zero or one. For $i=1,\ldots,p$ we combine $S_{2i-1}$ and $S_{2i}$ to produce $S'_i$. If $q=0$ then we move on to the next stage with the systems $S'_1,\ldots,S'_p$. Otherwise, $q=1$ and we combine $S'_p$ with $S_k$ to produce $S''_p$ and then proceed to the next stage with the systems $S'_1,\ldots,S'_{p-1},S''_p$.

We emerge from the overall procedure with an estimate of the form (7). If $\phi$ is minimized at each step, then we obtain an estimate of the largest singular value as in (8). The algorithm requires a small multiple of $n^2$ flops.

Finally, we remark that an $L_1$ divide and conquer estimator can obviously be formulated.

## 5. Test Results

The above condition estimators have been tested on numerous examples. In the $L_2$ case, we examined how well

E1 : Divide and Conquer

E2 : Look-behind (Algorithm 1) with weights $w_i = 1/t_{ii}$

E3 : Look-behind (Algorithm 1) with weights $w_i = 1$

could estimate the largest and smallest singular values of a given lower triangular matrix $T$.

Test 1.

- The lower triangular elements of $T$ were randomly selected from [-1,+1].
- 1000 examples were tried; 100 each for n = 5,10,15,20,25,30,35,40,45,50.
- The following table reports on the distribution of the "success measures" $q_n = \sigma_n/\hat{\sigma}_n$ and $q_1 = \hat{\sigma}_1/\sigma_1$.

| > | <= | E1 $q_n$ | E1 $q_1$ | E2 $q_n$ | E2 $q_1$ | E3 $q_n$ | E3 $q_1$ |
|---|---|---|---|---|---|---|---|
| .9 | 1.0 | 65.1% | 1.1% | 56.8% | 1.0% | 62.6% | 1.7% |
| .8 | .9 | 12.4% | 2.4% | 11.1% | 1.2% | 11.6% | 1.8% |
| .7 | .8 | 6.1% | 1.9% | 7.9% | 1.7% | 6.8% | 2.6% |
| .6 | .7 | 4.7% | 3.9% | 4.8% | 6.9% | 3.5% | 4.9% |
| .5 | .6 | 4.0% | 4.7% | 4.2% | 20.5% | 4.4% | 10.6% |
| .4 | .5 | 2.9% | 8.6% | 4.7% | 43.5% | 2.8% | 38.8% |
| .3 | .4 | 2.1% | 20.2% | 3.4% | 23.2% | 3.2% | 38.3% |
| .2 | .3 | 1.2% | 36.4% | 2.6% | 1.7% | 1.8% | 1.8% |
| .1 | .2 | 1.2% | 18.8% | 3.2% | .1% | 2.5% | .0% |
| .0 | .1 | .2% | 2.0 | 3.2% | .0% | .3% | .0% |

Comments

- We could not discern a correlation between the quality of the estimates and the condition of the matrix.
- The choice of weights in Algorithm 1 does not appear to be critical.
- We have no explanation why the estimates of $\sigma_1$ are consistently inferior to those for $\sigma_n$ .
- In no instance was either $q_1$ or $q_n$ less than .05.

Test 2.

- T generated by computing the QR-with-column-pivoting factorization of a square A whose entries are randomly selected from [-1,+1]. Specifically, the factorization AP = QT was computed where T is lower triangular and P is chosen to maximize $t_{kk}$ , k = n,...,2,1 in the k-th step.
- 1000 examples were tried; 100 each for n = 5,10,15,20,25,30,35,40,45,50.
- The following table reports on the distribution of $q_n = \sigma_n/\hat{\sigma}_n$ .

| > | <= | E1 : $q_n$ | E2 : $q_n$ | E3 : $q_n$ |
|---|---|---|---|---|
| .9 | 1.0 | 97.5% | 98.9% | 97.4% |
| .8 | .9 | 1.7% | .5% | 1.4% |
| .7 | .8 | .7% | .5% | .9% |
| .6 | .7 | .0% | .0% | .2% |
| .5 | .6 | .1% | .1% | .1% |
| .0 | .5 | .0% | .0% | .0% |

Comments

- By virtue of the column pivoting, $t_{kk}^2 \geq \sum_{i=1}^{k} t_{ij}^2$ for all $1 \leq j < k \leq n$ .
- The quality of the estimates for $\sigma_1$ for these matrices is essentially the same as for the Test 1 matrices and hence, unnecessary to report.
- In the majority of cases, $q_n > .99$ for all three methods.

The $L_1$ look-behind method (Algorithm 2 with $w_i = 1$) was also tested. A high degree of reliability was observed.

Test 3

- The lower triangular elements of T were selected randomly from [-1,+1]
- 250 examples were tried; 5 each for n = 1,2,...,50.
- The following table reports on the approximate distribution of $\hat{k}_1/k_1(T)$

| > | <= | $\hat{k}_1/k_1(T)$ |
|---|---|---|
| .99 | 1.00 | 78% |
| .50 | .99 | 6% |
| .10 | .50 | 12% |
| .05 | .10 | 4% |

## 6. Conclusions

It is important to interpret the above experimental results correctly. To begin with, they are just that--experimental results. They do not "prove" anything. However, they do suggest that our methods are at least as reliable as the LINPACK estimator which almost always produces $L_1$ estimates that are within a factor of 10 of the true condition. Should we therefore argue for the inclusion of one of our methods in LINPACK, especially since Cline [ 8 ] has produced an example upon which the LINPACK estimator fails?

This question focuses attention on the difficult problem of assessing condition estimation algorithms. Should one's enthusiasm for a 99% reliable method be diminished because of the existence of counter examples? If we believed this then we would not have presented the divide and conquer technique because it is easy to construct examples upon which it gives arbitrarily poor estimates! No, we must not now argue about whose condition estimator is "better." Instead, we must work to produce an efficient, provably reliable technique. We are personally excited by our approaches because the experimental results, e.g., Test 2, suggest to us that some rigorous result may well be possible for the case when $T$ is obtained via QR with column pivoting.

## Acknowledgements

We are grateful to Cleve Moler, David Gay, and Don Heller for sharing their thoughts on the problems considered in this paper.

## References

[1] A.K. Cline, C.B. Moler, G.W. Stewart, and J.H. Wilkinson,"An Estimate of the Condition Number of a Matrix," SIAM J. Numer.Anal.,16(1979)368-375.

[2] J.J. Dongarra, C.B. Moler, J.R.Bunch, and G.W.Stewart, LINPACK User's Guide, SIAM, Philadelphia, 1979.

[3] G.E.Forsythe and C.B. Moler, Computer Solution of Linear Algebraic Equations, Prentice-Hall, Englewood Cliffs, 1967.

[4] C.B.Moler, "Iterative Refinement in Floating Point," JACM, 14, 316-327.

[5] I. Karasalo, "A Criterion for Truncation of the QR Decomposition Algorithm for Singular Linear Least Squares Problem," BIT 14 , 1974, 156-166.

[6] D.P.O'Leary, "Estimating Matrix Condition Numbers," SISSC, 1, 1980, 205-209.

[7] B. Smith et al, EISPACK Guide, Springer Verlag, New York, 1974.

[8] A.K.Cline, "A Set of Counter Examples to the LINPACK Condition Estimator," Manuscript, Department of Computer Science, University of Texas, 1981.

# Demonstration of a Matrix Laboratory

Cleve Moler
Department of Computer Science
University of New Mexico
Albuquerque, New Mexico, 87131/USA

ABSTRACT. MATLAB is an interactive computer program that serves as a convenient "laboratory" for computations involving matrices. This paper gives three examples showing how MATLAB can be used in the classroom and in numerical analysis research.

MATLAB is an interactive computer program that serves as a convenient "laboratory" for computations involving matrices. It provides easy access to matrix software developed by the LINPACK and EISPACK projects [1-3]. The capabilities range from standard tasks such as solving simultaneous linear equations and inverting matrices, through symmetric and nonsymmetric eigenvalue problems, to fairly sophisticated matrix tools such as the singular value decomposition.

It is expected that one of MATLAB's primary uses will be in the classroom. It should be useful in introductory courses in applied linear algebra, as well as more advanced courses in numerical analysis, matrix theory, statistics and applications of matrices to other disciplines. In nonacademic settings, MATLAB can serve as a "desk calculator" for the quick solution of small problems involving matrices.

The program is written in Fortran and is designed to be readily installed under any operating system which permits interactive execution of Fortran programs. The resources required are fairly modest. There are less than 7000 lines of Fortran source code, including the LINPACK and EISPACK subroutines used. With proper use of overlays, it is possible run the system on a minicomputer with only 32K bytes of memory.

The size of the matrices that can be handled in MATLAB depends upon the amount of storage that is set aside when the system is compiled on a particular machine. We have found that an allocation of 5000 words for matrix elements is usually quite satisfactory. This provides room for several 20 by 20 matrices, for example. One implementation on a virtual memory system provides 100,000 elements. Since most of the algorithms used access memory in a sequential fashion, the large amount of allocated storage causes no difficulties.

In some ways, MATLAB resembles SPEAKEASY [4] and, to a lesser extent, APL. All are interactive terminal languages that ordinarily accept single-line commands or statements, process them immediately, and print the results. All have arrays or matrices as principal data types. But for MATLAB, the matrix is the only data type (although scalars, vectors and text are special cases), the underlying system is portable and requires fewer resources, and the supporting subroutines are more powerful and, in some cases, have better numerical properties.

Together, LINPACK and EISPACK represent the state of the art in software for matrix computation. EISPACK is a package of over 70 Fortran subroutines for various matrix eigenvalue computations that are based for the most part on Algol procedures published by Wilkinson, Reinsch and their colleagues [5]. LINPACK is a package of 40 Fortran subroutines (in each of four data types) for solving and analyzing simultaneous linear equations and related matrix problems. Since MATLAB is not primarily concerned with either execution time efficiency or storage savings, it ignores most of the special matrix properties that LINPACK and EISPACK subroutines use to advantage. Consequently, only 8 subroutines from LINPACK and 5 from EISPACK are actually involved.

This paper gives three extended examples involving linear equations, data fitting and partial differential equations that demonstrate MATLAB's capabilities. Further information on MATLAB, including the distribution tape and more complete documentation is available from the author.

## 1. Solving linear systems

When MATLAB is first entered, it prints a greeting, then prints the MATLAB prompt, <>, and waits for the user to type in a command from the terminal. For our first example, suppose we type the following line:

```
A = < 149 50 154; 537 180 546; 27 9 25 >
```

This enters a 3 by 3 matrix called A into the system. The angle brackets delineate the matrix and the semicolons indicate the ends of the rows. MATLAB responds by printing the following at the terminal:

```
A     =

   149.    50.   154.
   537.   180.   546.
    27.     9.    25.
```

It then gives another <> and waits again. We type

```
x = < 1/3 0 4*atan(1) >'
```

The prime at the end indicates that we want the row vector to be transposed and become a column vector. MATLAB responds

```
X     =

   0.3333
   0.
   3.1416
```

To multiply A times x and assign the result to b, we type

```
b = A*x
```

and obtain

```
B     =

   1.0d+03 *

   0.5335
   1.8943
   0.0875
```

There is a scale factor of $10^3$ which applies to the entire vector.

We are now ready to solve the system of simultaneous linear equations Ax = b . There are several ways in MATLAB to solve such systems. Two of the most common are produced by typing

```
y = inv(A)*b,  z = A\b
```

The vector y is obtained by first computing $A^{-1}$ and then multiplying b by the result. The backwards division sign indicates that the vector z is to be computed directly from A and b by Gaussian elimination without computing the inverse. MATLAB prints the results

```
Y     =

   0.3333
  -0.0000
   3.1416

Z     =

   0.3333
  -0.0000
   3.1416
```

We see that y and z both agree with the original x to the number of figures being printed. However, MATLAB has actually been doing its calculations with more accuracy than it is displaying. To use an output format which shows all the figures, we type

```
long, <x y z>
```

and obtain

```
ANS  =

   0.333333333333333   0.333333333345195   0.333333333343178
   0.                 -0.000000000005912  -0.000000000031322
   3.141592653589793   3.141592653590678   3.141592653590436
```

We see that, on this particular computer, both y and z differ from x in about the last 5 significant figures. This can be quantified by switching formats again and computing the actual error.

```
short e,  err = < norm(x-y), norm(x-z) >

ERR  =

   1.3282d-11   3.2839d-11
```

As expected, both y and z are accurate to about 11 significant figures. It turns out that y is slightly more accurate than z . However, there is another equally important measure of the "goodness" of an approximate solution to an equation -- how well does it satisfy the equation. In other words, what is the residual?

```
res = < norm(A*y-b), norm(A*z-b) >

RES  =

   6.0148d-09   8.5284d-14
```

We see that the residual of the solution obtained directly by Gaussian elimination is five orders of magnitude smaller than that of the solution obtained through multiplication by the inverse. Smallness of the residual is an important property of Gaussian elimination that can be established in general through Wilkinson's inverse error analysis. The five orders of magnitude is a consequence of the size of the condition number of this matrix,

```
cond(A)

ANS  =

   2.7585d+05
```

Of course, a condition number of $10^5$ is not a serious problem when computations are being done to 15 or 16 figures as we have been doing. However, MATLAB can simulate computations on computers with shorter word lengths. We now redo the example using an accuracy comparable to that of a computer with six hexadecimal figures in its floating point fraction.

```
chop(9), short, eps

EPS    =

   1.9073d-06

<>
y = inv(A)*b;  z = A\b;  <x y z>

ANS    =

   0.3333    0.2768    0.3252
   0.        0.2490    0.0259
   3.1416    3.1372    3.1411

<>
err = <norm(x-y), norm(x-z)>,  ...
res = <norm(A*y-b), norm(A*z-b)>

ERR    =

   0.2554    0.0271

RES    =

  12.5408    0.0010
```

We see that, on such a computer, both y and z have barely one significant figure (although, this time z is more accurate than y), but that the size of the residual makes z much more satisfactory than y for many applications.

This example not only illustrates some important points about roundoff errors in elementary matrix computations, but also shows how MATLAB can be used to demonstrate those points to students and other nonexperts.

## 2. Census example

Our next example involves predicting the population of the United States in 1980 using extrapolation of various fits to the census data from 1900 through 1970. There are eight observations, so we begin with the MATLAB statement

```
n = 8
```

The values of the dependent variable, the population in millions, can be entered with

```
y = < 75.995   91.972  105.711  123.203   ...
     131.669  150.697  179.323  203.212>'
```

In order to produce a reasonably scaled matrix, the independent variable, time, should be transformed from the interval [1900,1970] to [-1.00,0.75]. This can be accomplished directly with

```
t = -1.0:0.25:0.75
```

or in a fancier, but perhaps clearer, way with

```
t = 1900:10:1970;   t = (t - 1940*ones(t))/40
```

Either of these is equivalent to

```
t = <-1 -.75 -.50 -.25 0 .25 .50 .75>
```

The interpolating polynomial of degree n-1 involves an Vandermonde matrix of order n with elements that might be generated by

```
for i = 1:n, for j = 1:n, a(i,j) = t(i)**(j-1);
```

However, this results in an error caused by 0**0 when i = 5 and j = 1 . The preferable approach is

```
A = ones(n,n);
for i = 1:n, for j = 2:n, a(i,j) = t(i)*a(i,j-1);
```

Now the statement

```
cond(A)
```

produces the output

```
ANS   =

   1.1819E+03
```

which indicates that transformation of the time variable has resulted in a reasonably well conditioned matrix.

The statement

```
c = A\y
```

results in

```
C     =

  131.6690
   41.0406
  103.5396
  262.4535
 -326.0658
 -662.0814
  341.9022
  533.6373
```

These are the coefficients in the interpolating polynomial

$$c_1 + c_2 t + \cdots + c_n t^{n-1}$$

Our transformation of the time variable has resulted in $t = 1$ corresponding to the year 1980. Consequently, the extrapolated population is simply the sum of the coefficients. This can be computed by

```
p = sum(c)
```

The result is

```
P     =

  426.0950
```

which indicates a 1980 population of over 426 million. Clearly, using the seventh degree interpolating polynomial to extrapolate even a fairly short distance beyond the end of the data interval is not a good idea.

The coefficients in least squares fits by polynomials of lower degree can be computed using fewer than n columns of the matrix.

```
for k = 1:n, c = A(:,1:k)\y,  p = sum(c)
```

would produce the coefficients of these fits, as well as the resulting extrapolated population. If we do not want to print all the coefficients, we can simply generate a small table of populations predicted by polynomials of degrees zero through seven. We also compute the maximum deviation between the fitted and observed values.

```
for k = 1:n, X = A(:,1:k);  c = X\y;  ...
   d(k) = k-1;  p(k) = sum(c);  e(k) = norm(X*c-y,'inf');
<d, p, e>
```

The resulting output is

```
0   132.7227  70.4892
1   211.5101   9.8079
2   227.7744   5.0354
3   241.9574   3.8941
4   234.2814   4.0643
5   189.7310   2.5066
6   118.3025   1.6741
7   426.0950   0.0000
```

The zeroth degree fit, 132.7 million, is the result of fitting a constant to the data and is simply the average. The results obtained with polynomials of degree one through four all appear reasonable. The maximum deviation of the degree four fit is slightly greater than the degree three, even though the sum of the squares of the deviations is less. The coefficients of the highest powers in the fits of degree five and six turn out to be negative and the predicted populations of less than 200 million are probably unrealistic. The hopefully absurd prediction of the interpolating polynomial concludes the table.

We wish to emphasize that roundoff errors are not significant here. Nearly identical results would be obtained on other computers, or with other algorithms. The results simply indicate the difficulties associated with extrapolation of polynomial fits of even modest degree.

A stabilized fit by a seventh degree polynomial can be obtained using the pseudoinverse, but it requires a fairly delicate choice of a tolerance. The statement

```
s = svd(A)
```

produces the singular values

```
S     =

    3.4594
    2.2121
    1.0915
    0.4879
    0.1759
    0.0617
    0.0134
    0.0029
```

We see that the last three singular values are less than 0.1 , consequently, A can be approximately by a matrix of rank five with an error less than 0.1 . The Moore-Penrose pseudoinverse of this rank five matrix is obtained from the singular value decomposition with the following statements

```
c = pinv(A,0.1)*y, p = sum(c), e = norm(a*c-y,'inf')
```

The output is

```
C     =

 134.7972
  67.5055
  23.5523
   9.2834
   3.0174
   2.6503
  -2.8808
   3.2467

P     =

 241.1720

E     =

   3.9469
```

The resulting seventh degree polynomial has coefficients which are much smaller than those of the interpolating polynomial given earlier. The predicted population and the maximum deviation are reasonable. Any choice of the tolerance between the fifth and sixth singular values would produce the same results, but choices outside this range result in pseudoinverses of different rank and do not work as well.

The one term exponential approximation

$$y(t) = e^{pt}$$

can be transformed into a linear approximation by taking logarithms.

$$\log(y(t)) = \log k + pt$$
$$= c_1 + c_2 t$$

The following segment makes use of the fact that a function of a vector is the function applied to the individual components.

```
X = A(:,1:2);
c = X\log(y)
p = exp(sum(c))
e = norm(exp(X*c)-y,'inf')
```

The resulting output is

```
C     =

   4.9083
   0.5407
```

```
P     =

  232.5134

E     =

    4.9141
```

The predicted population and maximum deviation appear satisfactory and indicate that the exponential model is a reasonable one to consider.

As a curiousity, we return to the degree six polynomial. Since the coefficient of the high order term is negative and the value of the polynomial at t = 1 is positive, it must have a root at some value of t greater than one. The statements

```
X = A(:,1:7);
c = X\y;
c = c(7:-1:1);  //reverse the order of the coefficients
z = roots(c)
```

produce

```
Z     =

    1.1023-   0.0000*i
    0.3021+   0.7293*i
   -0.8790+   0.6536*i
   -1.2939-   0.0000*i
   -0.8790-   0.6536*i
    0.3021-   0.7293*i
```

There is only one real, positive root. The corresponding time on the original scale is

```
1940 + 40*real(z(1))

  =  1984.091
```

We conclude that the United States population should become zero early in February of 1984.

### 3. Partial differential equation example

Our next example is a boundary value problem for Laplace's equation. The underlying physical problem involves the conductivity of a medium with cylindrical inclusions and is considered by Keller and Sachs [6].

Find a function $u(x,y)$ satisfying Laplace's equation

$$u_{xx} + u_{yy} = 0$$

The domain is a unit square with a quarter circle of radius $\rho$ removed from one corner. There are Neumann conditions on the top and bottom edges and Dirichlet conditions on the remainder of the boundary.

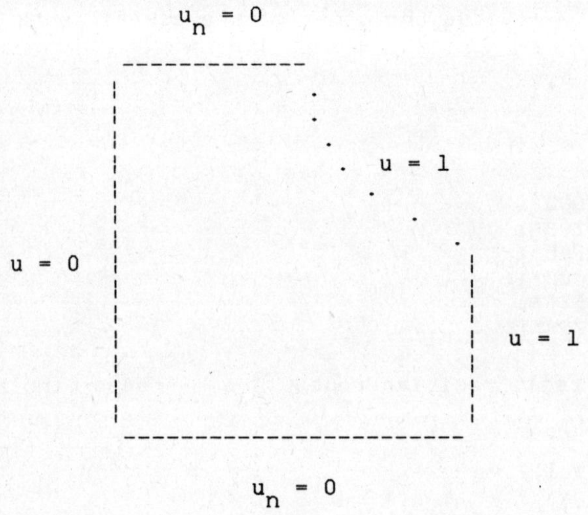


The effective conductivity of an medium is then given by the integral along the left edge,

$$\sigma = \int_0^1 u_n(0,y)\, dy$$

It is of interest to study the relation between the radius $\rho$ and the conductivity $\sigma$. In particular, as $\rho$ approaches one, $\sigma$ becomes infinite.

Keller and Sachs use a finite difference approximation. The following technique makes use of the fact that the equation is actually Laplace's equation and leads to a much smaller matrix problem to solve.

Consider an approximate solution of the form

$$u \approx \sum_{j=1}^{n} c_j r^{2j-1} \cos(2j-1)\theta$$

where $r,\theta$ are polar coordinates. The coefficients are to be determined. For any set of coefficients, this function already satisfies the differential equation because the basis functions are harmonic; it satisfies the normal derivative boundary condition on the bottom edge of the domain because we used $\cos\theta$ in preference to $\sin\theta$; and it satisfies the boundary condition on the left edge of the domain because we use only odd multiples of $\theta$.

The computational task is to find coefficients so that the boundary conditions on the remaining edges are satisfied as well as possible. To accomplish this, pick $m$ points $(r,\theta)$ on the remaining edges. It is desirable to have $m > n$ and in practice we usually choose $m$ to be two or three times as large as $n$. Typical values of $n$ are 10 or 20 and of $m$ are 20 to 60. An $m$ by $n$ matrix $A$ is generated. The $i,j$ element is the j-th basis function, or its normal derivative, evaluated at the i-th boundary point. A right hand side with $m$ components is also generated. In this example, the elements of the right hand side are either zero or one. The coefficients are then found by solving the overdetermined set of equations

$$Ac \approx b$$

in a least squares sense.

Once the coefficients have been determined, the approximate solution is defined everywhere on the domain. It is then possible to compute the effective conductivity $\sigma$. In fact, a very simple formula results,

$$\sigma = \sum_{j=1}^{n} (-1)^{j-1} c_j$$

To use MATLAB for this problem, the following "program" is first stored in the local computer file system, say under the name "PDE".

```
//Conductivity example.
//Parameters ---
   rho         //radius of cylindrical inclusion
   n           //number of terms in solution
   m           //number of boundary points
//initialize operation counter
   flop = <0 0>;
//initialize variables
   m1 = round(m/3);    //number of points on each straight edge
   m2 = m - m1;        //number of points with Dirichlet conditions
   pi = 4*atan(1);
//generate points in Cartesian coordinates
   //right hand edge
   for i = 1:m1, x(i) = 1; y(i) = (1-rho)*(i-1)/(m1-1);
   //top edge
   for i = m2+1:m, x(i) = (1-rho)*(m-i)/(m-m2-1); y(i) = 1;
   //circular edge
   for i = m1+1:m2, t = pi/2*(i-m1)/(m2-m1+1); ...
      x(i) = 1-rho*sin(t);  y(i) = 1-rho*cos(t);
//convert to polar coordinates
   for i = 1:m-1, th(i) = atan(y(i)/x(i));  ...
      r(i) = sqrt(x(i)**2+y(i)**2);
   th(m) = pi/2;  r(m) = 1;
//generate matrix
   //Dirichlet conditions
   for i = 1:m2, for j = 1:n, k = 2*j-1; ...
      a(i,j) = r(i)**k*cos(k*th(i));
   //Neumann conditions
   for i = m2+1:m, for j = 1:n, k = 2*j-1; ...
      a(i,j) = k*r(i)**(k-1)*sin((k-1)*th(i));
//generate right hand side
   for i = 1:m2, b(i) = 1;
   for i = m2+1:m, b(i) = 0;
//solve for coefficients
   c = A\b
//compute effective conductivity
   c(2:2:n) = -c(2:2:n);
   sigma = sum(c)
//output total operation count
   ops = flop(2)
```

The program can be used within MATLAB by setting the three parameters and then accessing the file. For example,

```
rho = .9;
n = 15;
m = 30;
exec('PDE')
```

The resulting output is

```
RHO    =

   .9000

N      =

  15.

M      =

  30.

C      =

   2.2275
  -2.2724
   1.1448
   0.1455
  -0.1678
  -0.0005
  -0.3785
   0.2299
   0.3228
  -0.2242
  -0.1311
   0.0924
   0.0310
  -0.0154
  -0.0038

SIGM   =

   5.0895

OPS    =

   16204.
```

A total of 16204 floating point operations were necessary to set up the matrix, solve for the coefficients and compute the conductivity. The operation count is roughly proportional to $mn^2$. The results obtained for $\sigma$ as a function of $\rho$ by this approach are essentially the same as those obtained by the finite difference technique of Keller and Sachs, but the computational effort involved is much less.

**Acknowledgement.**

Most of the work on MATLAB has been carried out at the University of New Mexico, where it is being supported by the National Science Foundation. Additional work has been done during visits to Stanford Linear Accelerator Center, Argonne National Laboratory and Los Alamos Scientific Laboratory, where support has been provided by NSF and the Department of Energy.

**References**

[1] J. J. Dongarra, J. R. Bunch, C. B. Moler and G. W. Stewart, LINPACK Users' Guide, Society for Industrial and Applied Mathematics, Philadelphia, 1979.

[2] B. T. Smith, J. M. Boyle, J. J. Dongarra, B. S. Garbow, Y. Ikebe, V. C. Klema, C. B. Moler, Matrix Eigensystem Routines -- EISPACK Guide, Lecture Notes in Computer Science, volume 6, second edition, Springer-Verlag, 1976.

[3] B. S. Garbow, J. M. Boyle, J. J. Dongarra, C. B. Moler, Matrix Eigensystem Routines -- EISPACK Guide Extension, Lecture Notes in Computer Science, volume 51, Springer-Verlag, 1977.

[4] S. Cohen and S. Piper, SPEAKEASY III Reference Manual, Speakeasy Computing Corp., Chicago, Ill., 1979.

[5] J. H. Wilkinson and C. Reinsch, Handbook for Automatic Computation, volume II, Linear Algebra, Springer-Verlag, 1971.

[6] H. B. Keller and D. Sachs, "Calculations of the Conductivity of a Medium Containing Cylindrical Inclusions", J. Applied Physics 35, 537-538, 1964.

# A FAST ALGORITHM FOR THE EUCLIDEAN DISTANCE LOCATION PROBLEM*

Michael L. Overton
Courant Institute of Mathematical Sciences
New York University
251 Mercer Street
New York, NY 10012, USA.

## 1. Introduction

Consider the problem of finding the point in an n-dimensional space for which a weighted sum of the Euclidean distances to m given points is minimized. We may write this as:

$$\text{P1:} \quad \min_{x \in R^n} \sum_{i=1}^{m} w_i \|x-b_i\|,$$

where the $\{w_i\}$ are positive real scalars, the $\{b_i\}$ are distinct vectors in $R^n$, and $\|x\|$ denotes the Euclidean norm $(\sum_{i=1}^{n} x_i^2)^{\frac{1}{2}}$. If $n = 1$ or $m \leq 2$ the problem is trivial, but if $n \geq 2$ and $m \geq 3$ it is more difficult to solve. Note that the solution, say $\overset{*}{x}$, may be at one of the points $b_i$, for example if one of the weights $w_i$ is much larger than the others. The problem was first posed by Fermat in the 17th century, in a form with $n = 2$, $m = 3$ and $w_1 = w_2 = w_3 = 1$. This case was solved geometrically by Torricelli. The more general version P1 is variously called the general Fermat problem, the Weber problem and the (weighted)(single-facility) location problem. The last name refers to an application where a single new "facility" is to be located in the position which minimizes the sum of distances to m existing "facilities." There is a large literature on location problems (e.g. see Francis and Goldstein (1974)).

Optimality conditions and an interesting dual formulation for P1 are given (for n=2) in Kuhn (1967). Many algorithms have been designed to solve P1. An iterative method solving a sequence of least squares problems was proposed by Weiszfeld (1937) and has since been advocated by Kuhn (1973) and Eckhardt (1975,1980). A recent paper of Voss and Eckhardt (1980) proves the linear convergence of a general variant of this algorithm. Schechter (1970) gives another method based on successive over-relaxation. Other methods include the subgradient algorithm of Chatelon, Hearn and Lowe (1978) and the convex programming method of Cordellier and Fiorot (1978). However, none of these methods have attractive rate of convergence properties; in all cases the rate of convergence is linear at best. In this paper, we present a new algo-

*This work was supported in part by the United States Department of Energy grant DE-AC02-76ER03077.

rithm which enjoys quadratic convergence under mild conditions. The importance of the new method is primarily that it can be generalized to give a fast algorithm for the much harder problem:

$$\text{P2:} \quad \min_{x \in R^n} \sum_{i=1}^{m} \| A_i^T x - b_i \|,$$

where the $\{A_i\}$ are $n \times d$ matrices, $n \geq d$, and the $\{b_i\}$ are d-vectors (see Section 6).

## 2. Gradient discontinuities and second derivatives.

Let us define

$$r_i(x) = (x - b_i) w_i \qquad i = 1, \ldots, m,$$

and

$$F(x) = \sum_{i=1}^{m} \| r_i(x) \| .$$

Notice that $F(x)$, the objective function of P1, is convex. By differentiating we obtain:

$$g_i(x) \equiv \nabla \| r_i(x) \| = \frac{w_i}{\| r_i(x) \|} r_i(x),$$

$$G_i(x) \equiv \nabla^2 \| r_i(x) \| = \frac{w_i^2}{\| r_i(x) \|} \left( I - \frac{r_i(x) r_i(x)^T}{\| r_i(x) \|^2} \right),$$

where I denotes the identity matrix, provided that $x \neq b_i$. Note that the Hessian term $G_i(x)$ is unbounded as $x \to b_i$, but that the gradient term $g_i(x)$ remains bounded, although it is of course discontinuous at $x = b_i$. Now let us define

$$g(x) = \sum_{\| r_i(x) \| \neq 0} g_i(x), \qquad G(x) = \sum_{\| r_i(x) \| \neq 0} G_i(x).$$

If the gradient and Hessian of $F(x)$ are defined it is clear that they are given by $g(x)$ and $G(x)$ respectively. The discontinuities in the gradient and consequent unboundedness of the Hessian are what cause the difficulty in solving P1.

## 3. Optimality conditions

Let us consider conditions for a point x to be a solution to P1. If $x \neq b_i$, $i = 1,\ldots,m$, then clearly a necessary and sufficient condition for x to be a solution is that the gradient g(x) is zero (by convexity).

Now suppose that $x = b_i$ for some i, and consider the change in F along a direction p. Clearly

$$\|r_i(x+p)\| = w_i\|p\|$$

for any direction $p \in R^n$. Thus the total directional derivative of F for any direction p is given by

$$F'(x;p) \quad \lim_{\alpha\to 0^+} \frac{F(x+\alpha p) - F(x)}{\alpha} = g(x)^T p + w_i\|p\|.$$

Therefore

$$F'(x;p) \geq \|p\| \,(w_i - \|g(x)\|).$$

by the Cauchy-Schwarz inequality. It follows that a necessary and sufficient condition for $x = b_i$ to be a minimum is that

$$\|g(x)\| \leq w_i. \qquad (1)$$

## 4. Outline of the algorithm.

In this section we outline how the algorithm works. Given a point $x^{(0)}$, generate a sequence of points $\{x^{(k)}\}$ as follows. If $x^{(k)}$ is not very close to $b_i$, $i = 1,\ldots,m$, then define

$$x^{(k+1)} = x^{(k)} + \alpha^{(k)} p^{(k)}$$

where the direction of search $p^{(k)}$ is obtained from the Newton equation

$$G(x^{(k)})p^{(k)} = -g(x^{(k)}) \qquad (2)$$

and the steplength $\alpha^{(k)}$ is chosen such that

$$F(x^{(k+1)}) < F(x^{(k)}).$$

Because F is convex, $G(x^{(k)})$ is always positive semi-definite and hence

$$g(x^{(k)})^T p^{(k)} = -p^{(k)T} G(x^{(k)}) p^{(k)} \leq 0.$$

Thus $p^{(k)}$ is always a descent direction for F unless $G(x^{(k)})$ is singular (which can be detected during the solution of the linear system (2)) or $g(x^{(k)})$ is zero (in which case $x^{(k)}$ is a solution). Under certain conditions on the <u>line search</u> algorithm used to define $\alpha^{(k)}$, it can be shown that the sequence of points $\{x^{(k)}\}$ has one of two properties. The first possibility is that $x^{(k)}$ stays bounded away from the $\{b_i\}$ and $g(x^{(k)})$ converges to zero as $k \to \infty$, in which case a solution is approximated. The second is that some iterate $x^{(\ell)}$ is generated which is close to some $b_i$, say with $\|r_i(x)\| \leq \varepsilon$. Once the latter situation is detected, $F(b_i)$ can be computed. If $F(b_i) \leq F(x^{(\ell)})$ then set $x^{(\ell+1)} = b_i$ and compute $g(b_i)(= \sum_{j \neq i} g_j(b_i))$. If (1) holds then a solution has been found. It is unlikely that these conditions will fail to hold given that a point close to $b_i$ was generated by the iteration, but if they do fail alternative steps can be taken. If $F(b_i) \leq F(x^{(\ell)})$ and (1) does not hold than set $p^{(\ell+1)} = -g(b_i)$, the steepest descent direction for F at $b_i$, and continue the iteration. If $F(b_i) > F(x^{(\ell)})$ then reduce $\varepsilon$ and contine the iteration with $p^{(\ell)}$ set to either the Newton step from (2) or a "regularized" step with the term $G_i$ omitted from $G(x^{(\ell)})$ in (2).

It follows from standard properties of Newton's method that if the solution $\overset{*}{x}$ is not equal to $b_i$, $i = 1,\ldots,m$, so that F is differentiable at $\overset{*}{x}$, then the asymptotic rate of convergence of $x^{(k)}$ to $\overset{*}{x}$ will be quadratic. What is not obvious is how quickly or slowly $x^{(k)}$ will approach the solution $\overset{*}{x}$ if $\overset{*}{x} = b_i$ for some i (a "non-smooth" minimum). If $x^{(k)}$ converges to such a point $b_i$, the Hessian term $G_i(x^{(k)})$ blows up and the total Hessian $G(x^{(k)})$ in (2) becomes extremely ill-conditioned. However it can be proved that this ill-conditioning, instead of creating difficulties, actually causes $x^{(k)}$ to converge quadratically to $b_i$, provided that an appropriate line search algorithm is used to compute $\alpha^{(k)}$. Consequently the fact that $x^{(k)}$ is converging to $b_i$ can be detected after just a few iterations. The essential feature of the line search is that it be capable of setting $\alpha^{(k)}$ to an estimate of either a "smooth" minimum (e.g. $\alpha^{(k)}=1$ if $x^{(k)}$ is near $\overset{*}{x} \neq b_i, i=1,\ldots,m$) or a "non-smooth" minimum (e.g. $\alpha^{(k)} = -\|x^{(k)}-b_i\|^2/(x^{(k)}-b_i)^T p^{(k)}$ if $x^{(k)}$ is near $\overset{*}{x}=b_i$). Details of the line search algorithm and the proof of quadratic convergence to a "non-smooth" solution under mild conditions are given (for the more general problem P2) in Overton (1981). The fact that superlinear (quadratic) convergence follows from solving increasingly ill-conditioned systems of equations is similar to the superlinear (cubic) convergence property of the Rayleigh quotient iteration for computing an eigenvector of a symmetric matrix (see Parlett (1980)).

5. Numerical examples.

The algorithm has been implemented in FORTRAN using a VAX-11/780 at the Courant Mathematics and Computing Laboratory. Double precision arithmetic, i.e. approximately 16 decimal digits of accuracy, was used throughout. The parameter $\varepsilon$ was set to $10^{-4}$. A number of test problems were solved, of which the following is a representative selection. In all cases the iterates converged to the solution and quadratic convergence was observed as predicted by the theory.

Example 1. (Non-smooth solution). $n = 2$, $m = 3$, $w_1 = w_3 = 1$, $w_2 = 2$.
$b_1 = [-1,0]^T$, $b_2 = [0,1]^T$, $b_3 = [1,0]^T$.
Solution $\overset{*}{x} = b_2$. Starting point $x^{(0)} = [3.0,2.0]^T$.

Results:

| k | $\|g(x^{(k)})\|$ | $\|r_2(x^{(k)})\|$ | $\alpha^{(k)}$ |
|---|---|---|---|
| 0 | 3.93 | 6.32 | $3.89\times10^{-2}$ |
| 1 | 1.18 | 1.00 | $4.98\times10^{-1}$ |
| 2 | 2.10 | $1.68\times10^{-1}$ | $3.58\times10^{-2}$ |
| 3 | $7.97\times10^{-1}$ | $3.81\times10^{-3}$ | $2.05\times10^{-3}$ |
| 4 | - | $1.75\times10^{-6}$ | - |
| 5 | 1.41 | 0.0 | - |

The column headed $\|r_2(x^{(k)})\|$ shows quadratic convergence to the non-smooth solution $b_2$ during iterations 1 through 4 (because of the ill-conditioning property). At the fifth iteration $x^{(5)}$ is set to $b_2$ and (1) is satisfied.

Example 2. (Smooth solution). $n = 2$, $m = 3$, $w_1 = w_2 = w_3 = 1$.
$b_1 = [-1,0]^T$, $b_2 = [0,1]^T$, $b_3 = [1,0]^T$.
Solution $\overset{*}{x} = [0.0,\ 0.577350]^T$.
Starting point $x^{(0)} = [3.0,\ 2.0]^T$.

Results:

| k | $\|g(x^{(k)})\|$ | $\alpha^{(k)}$ |
|---|---|---|
| 0 | 2.94 | $4.07\times10^{-2}$ |
| 1 | $4.60\times10^{-1}$ | 1.0 |
| 2 | $4.08\times10^{-1}$ | 1.0 |
| 3 | $8.69\times10^{-2}$ | 1.0 |
| 4 | $6.63\times10^{-3}$ | 1.0 |
| 5 | $2.23\times10^{-5}$ | 1.0 |
| 6 | $4.43\times10^{-10}$ | - |

The results show quadratic convergence to a smooth solution (because of standard properties of Newton's method).

Example 3 (Non-smooth solution).

This example is generated randomly, using the pseudo-random sequence:

$$\psi_0 = 7,\ \psi_{i+1} = (445\psi_i+1) \bmod 4096,\ i = 1,2,\ldots,$$

$$\overline{\psi}_i = \frac{\psi_i}{4095}\quad,\quad i = 1,2,\ldots\ .$$

The problem is defined by: $n = 3$, $m = 100$, $w_1 = 100$, $w_2 = w_3 = \ldots = w_m = 1$. The elements of $b_i$, $i = 1,\ldots,m$, are successively set to $\overline{\psi}_1, \overline{\psi}_2, \ldots,$ in the order $(b_1)_1,\ldots,(b_1)_n,(b_2)_1,\ldots,(b_m)_n$. The solution $\overset{*}{x}$ is $b_1$, and the initial point $x^{(0)}$ is $b_m$.

Results:

| k | $\lVert g(x^{(k)})\rVert$ | $\lVert r_1(x^{(k)})\rVert$ | $\alpha^{(k)}$ |
|---|---|---|---|
| 0 | $1.63\times10^{2}$ | $8.48\times10^{1}$ | $5.37\times10^{-3}$ |
| 1 | $1.01\times10^{2}$ | $2.38\times10^{1}$ | $4.10\times10^{-1}$ |
| 2 | $7.20\times10^{1}$ | 3.73 | $8.75\times10^{-2}$ |
| 3 | $7.33\times10^{1}$ | $1.25\times10^{-1}$ | $2.77\times10^{-3}$ |
| 4 | $6.83\times10^{1}$ | $1.38\times10^{-4}$ | $3.24\times10^{-6}$ |
| 5 | - | $1.67\times10^{-10}$ | - |
| 6 | $7.12\times10^{-1}$ | 0.0 | - |

Here $x^{(0)} = b_m$ and (1) does not hold so a steepest descent step is taken at the first iteration. The iterates then converge quadratically to the solution as in Example 1. At the sixth iteration $x^{(6)}$ is set to $b_1$ and (1) is satisfied.

## 6. Concluding remarks

The results of Section 5 illustrate that the new algorithm gives highly accurate solutions in just a few iterations, because of the quadratic convergence property which holds whether or not F is differentiable at the solution. Naturally, P1 can be solved by simpler methods. For example, define i by $F(b_i)=\min_{1\le j\le m} F(b_j)$; then if $\lVert g(b_i)\rVert \le w_i$, $b_i$ is a solution and otherwise a Newton iteration converging to a "smooth" solution could be used. However the importance of the new algorithm is that it can be generalized to give a quadratically convergent method to solve the more difficult problem P2, where <u>many</u> terms $\{A_i^T x-b_i\}$ could be zero at the solution. This method is described in Overton (1981). A related algorithm for P2 is given by Calamai and Conn (1980).

Acknowledgment. The author would like to thank Gene H. Golub for bringing the subject of this paper to his attention.

## References

P.H. Calamai and A.R. Conn (1980). A stable algorithm for solving the multifacility location problem involving Euclidean distances, SIAM J. Scient. and Stat. Comp. 1, pp. 512-526.

J.A. Chatelon, D.W. Hearn and T.J. Lowe (1978). A subgradient algorithm for certain minimax and minisum problems, Math. Prog. 15, pp. 130-145.

F. Cordellier and J. Ch. Fiorot (1978). On the Fermat-Weber problem with convex cost functionals, Math. Prog. 14, pp. 295-311.

U. Eckhardt (1975). On an optimization problem related to minimal surfaces with obstacles, in R. Bulirsch, W. Oettli and J. Stoer, eds., Optimization and Optimal Control, Lecture Notes in Mathematics 477, Springer-Verlag, Berlin and New York.

U. Eckhardt (1980). Weber's problem and Weiszfeld's algorithm in general spaces, Math. Prog. 18, pp. 186-196.

R.L. Francis, and J.M. Goldstein (1974). Location theory: a selective bibliography, Operations Research 22, pp. 400-410.

H.W. Kuhn (1967). On a pair of dual nonlinear programs, in Nonlinear Programming (J. Abadie, ed.), North-Holland, Amsterdam, pp. 38-54.

H.W. Kuhn (1973). A note on Fermat's problem, Math. Prog. 4, pp. 98-107.

M.L. Overton (1981). A quadratically convergent algorithm for minimizing a sum of Euclidean norms, Computer Science Dept. Report 030, Courant Institute of Mathematical Sciences, New York University.

B.N. Parlett (1980). The Symmetric Eigenvalue Problem, Prentice-Hall, Englewood Cliffs, N.J.

S. Schechter (1970). Minimization of a convex function by relaxation, in Integer and Nonlinear Programming (J. Abadie, ed.), North-Holland, Amsterdam and London.

H. Voss and U. Eckhardt (1980). Linear convergence of generalized Weiszfeld's method, Computing 25, pp. 243-251.

E. Weiszfeld (1937). Sur le point par lequel la somme des distances de n points donnés est minimum, Tohoku Mathematics J. 43, pp. 355-386.

# DISCRETE PRESSURE EQUATIONS IN INCOMPRESSIBLE FLOW PROBLEMS

E. L. WACHSPRESS
General Electric Company
KNOLLS ATOMIC POWER LABORATORY
Schenectady, New York, 12301/USA
Operated for the United States
Department of Energy
Contract DE-AC12-76SN00052

ABSTRACT

Consistent integration of flow divergence and pressure gradient terms over interlocking mesh boxes preserves properties that enhance stability and convergence of discretized Navier-Stokes equations over regions partitioned with isoparametric quadrilaterals.

A standard problem in the calculus of variations is to find w in a trial space W such that the functional (w,Hw)-2(w,f) is minimized, subject to the constraint Bw-g=0. Here H (self-adjoint) and B are prescribed operators defined over W while f and g are given functions. One may introduce a Lagrangian-multiplier function, h, and modify the functional to include the constraint:

$$F(w,h) = (w,Hw) - 2(w,f) + 2(h, Bw - g).$$

The first variation of F with respect to w is zero when

$$Hw + B^*h - f = 0$$ (where * denotes the adjoint operator).

The first variation of W with respect to h is zero when

$$Bw - g = 0.$$

The function h does not appear in the initial statement of the problem. Note that at the solution point, $w_o$, the last term in the functional vanishes. Nevertheless, h plays a crucial role in enforcing the constraint. When F is an energy functional, the variation of F with respect to w yields a force balance equation. This suggests that the term B*h is a force introduced by the constraint. The analysis may be applied to non-self-adjoint H, in which case one may introduce an adjoint function, w*. The Navier-Stokes equations are a particular form of the above principle. The operator H is $\nabla \cdot (\rho\vec{v} - \mu\nabla)$ and the constraint is the continuity equation $\nabla \cdot \rho\vec{v} = 0$. Here, ρ is density, μ is viscosity, and $\vec{v}$ is fluid velocity. The Lagrangian-multiplier is now called "pressure". The adjoint of the gradient operator is the negative of the divergence operator. Non-linearity evidenced by the $\vec{v}$-dependence of H leads to enormous complexities in analysis. Numerical solution is often accomplished with a linearized system in which previously estimated velocities are used in H. In this discussion, this known value will be denoted by $\vec{v}_o$ and only the linearized equations will be examined. The stationarity equations with respect to velocity and pressure are the Navier-Stokes equations governing steady flow:

$$\nabla \cdot (\rho \, \bar{v}_0 \bar{v} - \mu \nabla \bar{v}) + \nabla p - \bar{f} = \bar{0} \tag{1.1}$$

$$\bar{\nabla} . \, \rho \bar{v} = 0 \tag{1.2}$$

Equation 1.1 is the force balance and Equation 1.2 is the mass balance. Only constant density (incompressible flow) will be treated here. The pressure in the fluid adjusts instantaneously to conserve mass throughout the fluid. Pressure is thus an "elliptic" variable. This suggests that in numerical solution pressures should be solved by an implicit procedure subject to boundary conditions around the entire domain of interest and driven by local mass imbalance that may be introduced by incorrect velocity estimates during the course of solution.

The discrete form of Equations 1, whether obtained by finite-difference or element techniques, is

$$M \, \underline{v} + A \, \underline{p} - \underline{f} = \underline{0} \tag{2.1}$$

$$B \, \underline{v} - \underline{g} = \underline{0} \tag{2.2}$$

where the components of $\underline{v}$ are velocity components at nodes, the components of $\underline{p}$ are nodal pressures, and $\underline{g}$ arises from inlet and outlet flow boundary conditions.

Matrix A in Equation 2.1 is a discrete representation of the gradient operator. The gradient of a constant is zero. Let $\underline{h}^T=(1,1,\ldots,1)$ be a vector with as many components as $\underline{p}$. Then $A\underline{h}=\underline{0}$ for any reasonable discretization of the gradient. It would be beneficial if the null space of A were only of dimension one, but this is sometimes not the case, as will be discussed in greater depth subsequently. Suppose matrix M in (2.1) is nonsingular . To guarantee a nonsingular M, one may resort to "upwinding" of the transport term and to other less-advertised devices. As element size shrinks in the discretization, the viscosity term dominates the transport term and M becomes diagonally dominant in commonly used schemes. In any case, the existence of some solution to Equations 2 depends on M being non-singular. Solving (2.1) for $\underline{v}$ and substituting in (2.2), one obtains the pressure equation

$$- B M^{-1} A \underline{p} = \frac{1}{\rho} \underline{g} - B M^{-1} \underline{f}. \quad (3)$$

Since $A \underline{h}=\underline{0}$ for the nontrivial vector $\underline{h}$, $BM^{-1}A$ must be singular. The net gain of fluid mass over the region of concern is $\rho \underline{h}^T \underline{g}$, and, in the absence of internal sources or sinks, this must vanish. Matrix B is the discrete representation of the divergence operator. A reasonable requirement is that $\underline{h}^T B=\underline{0}^T$. Then the right hand side of Equation 3 is orthogonal to the known null-space component of the left-hand side of (3). Equation 3 may have a solution. Even when the null space of $BM^{-1}A$ is of dimension greater than one, the right-hand side of (3) can be generated to admit a solution; this will be clarified subsequently. Let $X^+$ denote the generalized inverse of X and let $\underline{q}$ be an arbitrary vector in the null space of $BM^{-1}A$; then a solution to (3) may be written as:

$$\underline{p} = (-BM^{-1}A)^+ (\frac{1}{\rho}\underline{g} - BM^{-1}\underline{f}) + \underline{q}. \quad (4)$$

Having found $\underline{p}$, one may substitute into (2.1) to obtain

$$\underline{v} = M^{-1}(\underline{f} - A\underline{p}). \quad (5)$$

Note that when the null space of $BM^{-1}A$ is the null space of A, $A\underline{q}=\underline{0}$ and $\underline{v}$ in (5) is unique even though the pressure is nonunique. When $\underline{q}$ is a multiple of $\underline{h}$, the arbitrariness in pressure is removed by normalizing to a prescribed value for pressure at a specified node.

In commonly used methods for solving (2), one replaces M in $BM^{-1}A$ by the diagonal component, D, of matrix M and obtains a "pressure correction" equation of the form

$$\underline{p}'_t = (-BD^{-1}A)^+ \underline{s}_t \quad (6)$$

where $\underline{s}_t$ is a function of the latest velocity estimate, $\underline{v}_t$. Then one computes the new velocity estimate from

$$\underline{v}_{t+1} = M^{-1}(\underline{f} - A\underline{p}_{t+1}) \quad (7)$$

with $\underline{p}_{t+1}=\underline{p}_t+\underline{p}'_t$. In practice, (7) is not used with the exact $M^{-1}$; some iterative approximation to $\underline{v}_{t+1}$ is used. The velocity updating equation must be underrelaxed for the iteration to converge. This is described in Reference 1.

An alternative derivation of a pressure equation is based on taking the divergence of (1.1) with application of (1.2) to reduce the velocity terms. For example, when viscosity is constant,

$$\nabla \cdot (\nabla \cdot \mu \nabla \bar{v}) = \mu \nabla \cdot \nabla^2 \bar{v} = \mu \nabla^2 (\nabla \cdot \bar{v}) = 0 .$$

The pressure equation in this formulation is a Poisson equation.

As mentioned previously, the pressure equation should be elliptic. Matrix (-BA) is a discrete approximation to the positive semidefinite operator $-\nabla^2$. If the entire null space of A is $\underline{h}$ and if $B=-A^T$, then matrix $(-BD^{-1}A)$ is symmetric and positive semidefinite for D positive diagonal. The null space of $BD^{-1}A$ is $\underline{h}$.

In some discretization procedures there are more pressure components (order of the pressure vector) than there are cells over which mass conservation is imposed. The physical role of pressure is to force mass conservation. By allowing more degrees of freedom in the pressure vector, one inadvertently admits pressure vectors other than $\underline{h}$ in the null space of A. These vector components do not affect the velocity. However, they introduce the spurious pressures that are observed in some computations. This is known as "checkerboarding" because these modes are often oscillatory. To avoid checkerboarding, one must take care to introduce just enough pressure components to achieve mass balances around all elements. An interesting account of spurious pressure modes appears in Reference 2. Much else has been written on checkerboarding and means for smoothing out such spurious modes.

Consider a grid of interlocking velocity and pressure boxes. The divergence term may be converted to a line integral around each pressure box (Sketch 1). In the force equation, the pressure gradient approximated at a velocity node may be multiplied by the area of the box around this node to approximate the pressure term (Sketch 2).

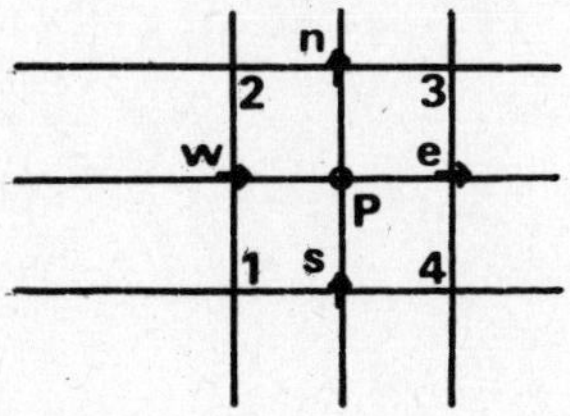


Sketch 1. Pressure box at P

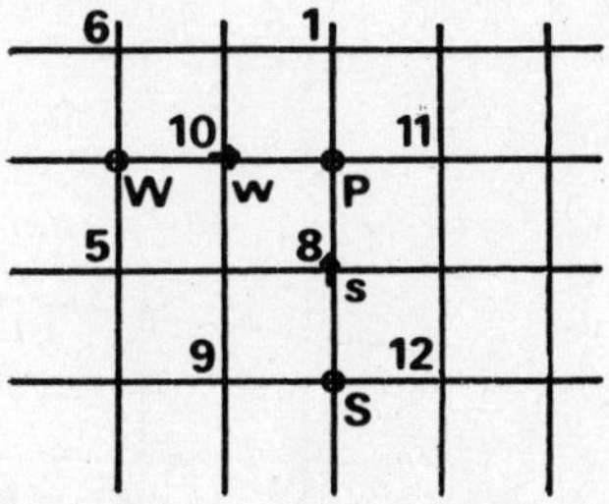


Sketch 2. Velocity boxes at w and s

$$\iint_{p\text{-box}} \nabla \cdot \rho \bar{v} \, dx dy = \int_{\partial p \, (1,2,3,4)} \bar{n} \cdot \rho \bar{v} \, d\ell \doteq \rho [ u_e - u_w)(y_2 - y_1) + (v_n - v_s)(x_3 - x_2)] ,$$

$$\iint\limits_{\substack{\text{w-box}\\(5,6,7,8)}} p_x dxdy \doteq \frac{(p_p - p_w)}{(x_8 - x_5)} (x_8 - x_5)(y_6 - y_5)$$

$$\iint\limits_{\substack{\text{s-box}\\(9,10,11,12)}} p_y dxdy \doteq \frac{(p_p - p_s)}{(y_{10} - y_9)} (x_{12} - x_9)(y_{10} - y_9)$$

It seems inconsistent to convert the divergence integral to a line integral while leaving the gradient as an area integral. For this rectangular grid, the two approximations are identical. Let $d_p$ be the diagonal coefficient of the velocity component at p in the force equation at p. Then the pressure equation at P is obtained from:

$$\begin{aligned} u_e &= \frac{1}{d_e}(p_P - p_E)(y_n - y_s) + \ldots \\ u_w &= \frac{1}{d_w}(p_W - p_P)(y_n - y_s) + \ldots \\ v_n &= \frac{1}{d_n}(p_P - p_N)(x_e - x_w) + \ldots \\ v_s &= \frac{1}{d_s}(p_S - p_P)(x_e - x_w) + \ldots \end{aligned} \tag{8.1}$$

$$\begin{aligned} (\underline{S}_t)_p &= \rho[(y_n - y_s)(u_e - u_w) + (x_e - x_w)(v_n - v_s)]_t \\ &= \rho\{(y_n - y_s)^2[\frac{1}{d_e}(p_P' - p_E') + \frac{1}{d_w}(p_P' - p_W')] \\ &\quad + (x_e - x_w)^2[\frac{1}{d_n}(p_P' - p_N') + \frac{1}{d_s}(p_P' - p_S')]\}. \end{aligned} \tag{8.2}$$

It is apparent that this leads to a symmetric positive semidefinite coefficient matrix. The motivation for converting the divergence area-integral into a line-integral while retaining the area-integral form for the pressure gradient is the desire for symmetric pressure-correction equations over nonrectangular grids. Consider, for example, the trapezoid in Sketch 3.

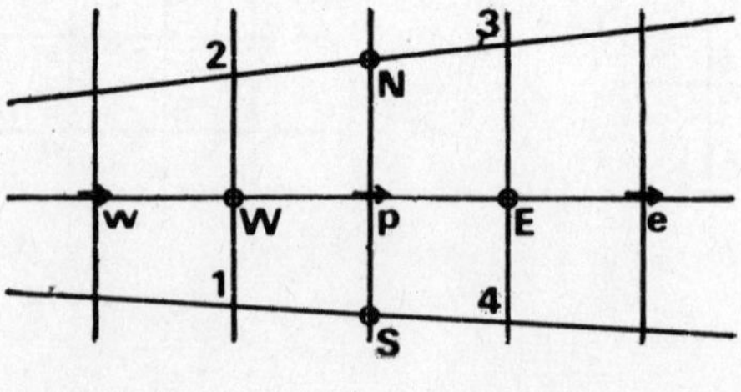


Sketch 3. A nonrectangular grid

The flow across side (S;p;N) is approximated by

$$\int_S^N \rho\bar{n} \cdot \bar{v}\, d\ell \doteq \rho\, u_p(y_N - y_S). \qquad (9)$$

If the pressure gradient term were treated as a line integral also, then in the force balance around Velocity Node p the pressure term would be

$$\vec{i}\cdot\iint_{p\text{-box}} \nabla p\, dx dy = \oint_{\partial p} \vec{i}\cdot\vec{n} p\, d\ell \qquad (10)$$

$$\doteq p_E(y_3 - y_4) + p_N(y_2 - y_3) - p_N(y_2 - y_1) - p_S(y_1 - y_4)$$

The coefficient of $p_W$ in the pressure-correction equation at E would be

$$\rho/d_p\ (y_N - y_S)(y_1 - y_2)$$

while the coefficient of $p_E$ in the pressure-correction equation at W would be

$$\rho/d_p\ (y_N - y_S)(y_3 - y_4).$$

Now consider the effect of multiplying a numerical approximation of the pressure gradient at Velocity Node p by the area of the box centered at p:

$$\vec{i}\cdot\iint_{p\text{-box}} \nabla p\, dx dy \doteq \frac{p_E - p_W}{x_E - x_W}(x_E - x_W)\frac{(y_2 - y_1) + (y_3 - y_4)}{2} \qquad (11)$$

and the coefficient of $p_E$ in the pressure-correction equation at W is equal to the coefficient of $p_W$ in the equation at E:

$$\rho/d_p\ (y_N - y_S)\frac{[(y_1 - y_2) + (y_3 - y_4)]}{2}. \qquad (12)$$

It will now be shown that symmetry is preserved even for nine-node isoparametric elements. A vector momentum equation may now be generated for each velocity-node and both components of velocity computed. The configuration of concern here is shown in Sketch 4.

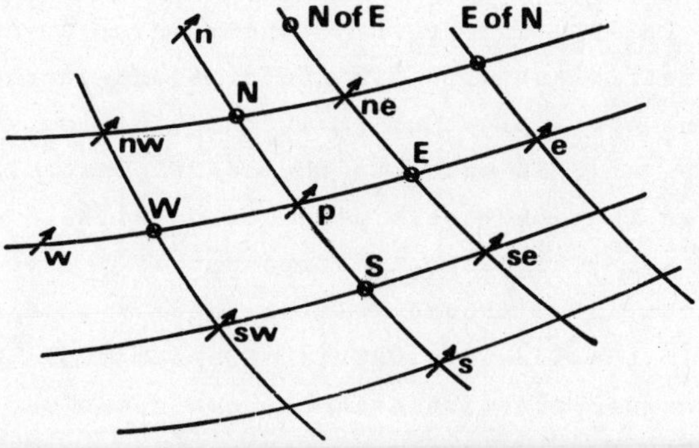


Sketch 4. Isoparametric element

The flow across side (S;p;N) of pressure boxes centered at W and E is

$$\rho[(y_N-y_S)\,u_p-(x_N-x_S)v_p]. \tag{13}$$

The flow across side (W;p;E) of the pressure boxes centered at S and N is

$$\rho[(x_E-x_W)v_p-(y_E-y_W)u_p]. \tag{14}$$

The pressure term in the momentum equation at p may be obtained by use of the one-node isoparametric quadrature formula

$$\begin{aligned}\iint_{\text{p-box}} dxdy\ f(x,y) &= \int_{-1}^{1}\int_{-1}^{1} d\xi\ d\eta\ J\left(\frac{x,y}{\xi,\eta}\right) f'(\xi,\eta)\\ &\doteq 4\ J(\xi=0,\eta=0)\ f'(\xi=0,\eta=0)\end{aligned} \tag{15}$$

where f' denotes evaluation of f(x,y) at the point in the $\xi,\eta$ plane into which (x,y) is mapped. The approximation to the gradient of pressure at Node p where $(\xi,\eta)=(0,0)]$ may be chosen as

$$\begin{bmatrix} p_x \\ p_y \end{bmatrix} = \begin{bmatrix} \xi_x & \eta_x \\ \xi_y & \eta_y \end{bmatrix}\begin{bmatrix} p_\xi \\ p_\eta \end{bmatrix} = \frac{1}{J}\begin{bmatrix} y_\eta & -y_\xi \\ -x_\eta & x_\xi \end{bmatrix}\begin{bmatrix} p_\xi \\ p_\eta \end{bmatrix}$$

$$\begin{bmatrix} p_x \\ p_y \end{bmatrix}_{0,0} = \frac{1}{4J(0,0)}\begin{bmatrix} (y_N-y_S) & (y_W-y_E) \\ (x_S-x_N) & (x_E-x_W) \end{bmatrix}\begin{bmatrix} (p_E-p_W) \\ (p_N-p_S) \end{bmatrix}. \tag{16}$$

From (15) and (16), one obtains

$$\iint_{\text{p-box}} \begin{bmatrix} p_x \\ p_y \end{bmatrix} dxdy \doteq \begin{bmatrix} (y_N-y_S) & (y_W-y_E) \\ (x_S-x_N) & (x_E-x_W) \end{bmatrix}\begin{bmatrix} (p_E-p_W) \\ (p_N-p_S) \end{bmatrix}. \tag{17}$$

Since only $(p_E-p_W)$ and $(p_N-p_S)$ appear in (17), one can easily deduce that the coefficient of $p_E$ in the pressure-correction equation at W is equal to the coefficient of $p_W$ in the equation at E and that the coefficient of $p_N$ at S is equal to the coefficient of $p_S$ at N. It remains to be shown that the coefficient of $p_N$ at E is equal to the coefficient of $p_E$ at N. The diagonal component of the vector momentum equation (force balance) around Velocity Node p is $d_pI$, where $d_p$ is positive and I is the identity matrix of order two. The coefficient of $p_N$ in the pressure equation at E has contributions from the flow across sides (S;p;N) and (N;ne; E of N). The coefficient of $p_E$ in the equation at N has contributions from flow across sides (W;p;E) and

(E;ne;N of E). It will now be shown that the contributions from flow across (S;p;N) and (W;p;E) are the same. That the other contributions must also be the same follows from geometric similarity. From (13) and (17), the contribution from flow across (S;p;N) to the coefficient of $p_N$ at E is

$$\rho/d_p \left[ \underbrace{(y_N - y_S)}_{u_p\text{-coef. in (13)}} \; \underbrace{(y_W - y_E)}_{p_N\text{-coef. in } p_x \text{ term in (17)}} \; - \; \underbrace{(x_N - x_S)}_{v_p\text{-coef. in (13)}} \; \underbrace{(x_E - x_W)}_{p_N\text{-coef. in } p_y \text{ term in (17)}} \right] \qquad (18)$$

From (14) and (17), the contribution from flow across (W;p;E) to the coefficient of $p_E$ at N is

$$\rho/d_p \left[ \underbrace{(x_E - x_W)}_{v_p\text{-coef. in (14)}} \; \underbrace{(x_S - x_N)}_{p_E\text{-coef. in } p_y \text{ term in (17)}} \; - \; \underbrace{(y_E - y_W)}_{u_p\text{-coef. in (14)}} \; \underbrace{(y_N - y_S)}_{p_E\text{- coef. in } p_x \text{ term in (17)}} \right] \qquad (19)$$

It is seen that these two coefficients are the same. Detailed examination of all coefficients reveals that the pressure-correction equations are symmetric.

The equations are not particularly complicated. One feature of this approach is that when the grid becomes rectangular the 5-point equations are recovered. In general, nine-point difference stars result at interior pressure nodes (Sketch 5).

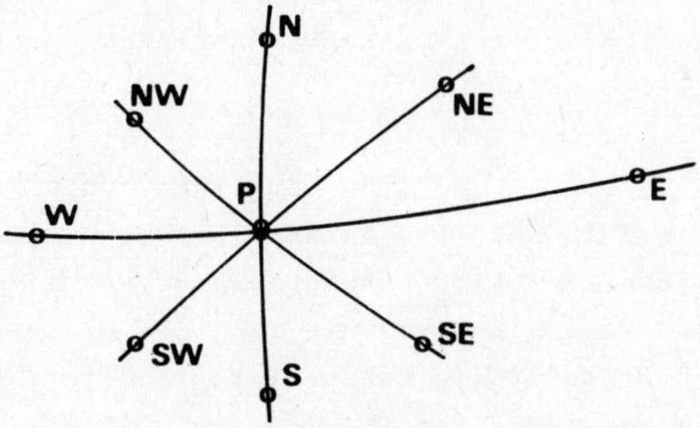


Sketch 5. Nine point pressure star

These equations may be considered in two sets, one for nodes on even columns and one for nodes on odd columns. The boxes for each set cover the entire grid without overlap. A different additive normalization may be applied to each set. The null space of the pressure coefficient matrix is thus at least of order two. This nonphysical peculiarity of the pressure equations is a result of the discretization scheme. It should not be confused with checkerboarding. One may adjust the additive normalizations so that the even and odd pressures are consistent at some reference point where one chooses the odd-node pressure as the average of neighboring even-node pressures. There is a flow balance equation for each pressure node. There are no spurious modes. It is true that pressures in a set (odd or even) are coupled more tightly to one another than to pressures in the other set. Discretization error caused by not choosing small enough elements for a given problem manifests itself in misalignment of odd and even pressures.

The righthand side of the discrete system must be orthogonal to the nullspace of the coefficient matrix if the pressure equations are to be well posed. The sum of the components of the source vector for the odd set and for the even set must each vanish. In rectangular geometry, only one set exists and outlet flow may be adjusted just prior to the pressure-correction calculation to maintain exact balance between inlet and outlet flow. For curved geometry this same balance may be forced. The sum of both even and odd source components is equal to the net gain in mass which can be made zero by adjustment of outlet flow just prior to the pressure calculation. The nine-point pressure stars yield a system that is efficiently solved by either two-line successive overrelaxation or two-line Chebyshev iteration. Overrelaxation has been used thus far but the Chebyshev approach is envisioned as more suitable for vector computation.

The pressure terms in the force equations could just as well have been determined directly from the balance equations as the transpose of the divergence operator times the pressure vector; this is quite reasonable when one views the pressure function as a Lagrangian multiplier for the mass conservation constraint. However, the approach followed here gives greater insight into the consistent approximation of divergence and gradient operators over interlocking velocity and pressure grids.

Alternative finite-element-type derivations automatically yield symmetric-positive-semidefinite pressure matrices. More complicated pressure stars often result from such methods. Also, the equations do not reduce to the five-point equations when the elements are rectangles. It is not clear that reduction to five-point equations is best.

Some considerations should be given to force-conservation. When the pressure terms in the force equation is converted to a line integral, the pressure forces are treated conservatively. In the symmetric formulation the equations are not conservative. Finite-element equations based on Galerkin-weighting are not necessarily conservative either. It seems that one may choose either conservative equations or symmetric equations but that there is no obvious way for obtaining both for the general geometry.

## REFERENCES

1.- Wachspress, E.L., "The numerical solution of turbulent flow problems in general geometry", "Numerical Analysis". Proceedings of the 8th Biennial Conference held in Dundee, Scotland, June 26-29, Edited by G.A. Watson. Lecture Notes in Mathematics. No. 773, pp. 146-163, Springer-Verlag, 1980. (The conservative pressure equations were derived in this reference).

2.- Sani, R.L., P.M. Gresho, R.L. Lee, and D.F. Griffiths. "The cause and cure(?) of the spurious pressures generated by certain FEM solutions of the incompressible Navier-Stokes equations". UCRL-84867. (August 1980 preprint of a paper to be published in the Intl. J. for Numerical Methods in Fluids).

# STANDING WAVES IN DIFFUSIVE REACTING SYSTEMS

SUSANA KAUFMANN
ANTONIO MONTALVO
IIMAS-UNAM
México 20, D.F.

1.- Introduction. During the last ten years some attention has been given in the literature to the problem of instabilities and sustained oscillations in systems with chemical reaction and diffusion. Among the published literature we might pointout the paper by Turing [1] and the book by Prigogine and Glansdorff [2], and more recently the book by Thom. In his paper, Turing established the basis of morphogenesis as a possible consequence of chemical reaction processes; Prigogine and Gansdorff gave a unified approach to a sort of phenomena exhibiting similar thermodynamic characteristics of the process described by Turing. Finally, Thom presented a broad class of phenomena, including those by Turing and Prigogine, but using a purely topological approach [3].

In this paper we analyse one of the examples described in [2]. For this problem we derive necessary and sufficient conditions for an homogeneous steady state bifurcates into an unstable one that might develop either spatial patterns or cyclic time behaviour. The particular final state of the system is predicted according to Hopf's bifurcation theory [4]. Finally, the stability conditions as well as the ultimate state of the system are verified by the integration forward in time of the equations describing the phenomenon.

2.- Problem Statement

Consider the hypothetical reacting system

$$A \underset{k_5}{\overset{k_1}{\rightleftarrows}} U \qquad (A)$$

$$2U + V \underset{k_6}{\overset{k_2}{\rightleftarrows}} 3U \qquad (B)$$

$$B + U \underset{k_7}{\overset{k_3}{\rightleftarrows}} V + D \qquad (C)$$

$$U \underset{k_8}{\overset{k_4}{\rightleftharpoons}} E \qquad (D)$$

Let's make the following assumptions regarding the characteristics of the above reacting system and its environment

a) the mass action law for the rates of reaction is valid

b) the system is open to components A, B, D, and E, that is, their concentration remains constant in the entire system

c) the system remains closed to components U and V

d) the diffusion coefficients for components U and V are constant.

Under the above assumptions the equations describing the time and space behavior of u and v are:

$$\frac{\partial u}{\partial t} = Du\nabla^2 U + R^u(u,v) \qquad (2.1a)$$

$$\frac{\partial v}{\partial t} = Dv\nabla^2 V + R^v(u,v) \qquad (2.1b)$$

$$(x,y,t) \in \Omega \times \tau = \{(o,Lx)x(o,Ly)\}\ x(o,T]$$

subject to the boundary conditions

$$\frac{\partial u}{\partial \eta} = \frac{\partial v}{\partial \eta} = 0 \qquad (x,y) \in \partial\,\Omega,\ t > 0 \qquad (2.1c)$$

In the above equations

$$R^u(u,v)=k_1 A+k_2 u^2 v-k_3 Bu-k_4 u-k_5 u-k_6 u^3+k_7 Dv\ k_8 E \qquad (2.2a)$$

$$R^v(u,v)=-k_2 u^2 v + k_3 Bu + k_6 u^3 + k_7 Dv$$

We now raise the following questions:

1) Under what conditions an homogeneous steady state becomes unstable?

2) What will be the time behavior of the system, if a given state $(u_0,v_0)$ satisfies (a)?

A basic requirement for a given state $(u_0,v_0)$ to be unstable is for the concentrations A, B, D, and E not to satisfy the chemical (thermodynamic) equilibrium. In the analysis to follow, we will assume that this requirement is satisfied.

In order to answer the first equation above stated, we carry out the usual frequency analysis on the linearized version of equations (2.1), in the neighborhood of $(u_0,v_0)$, that is

$$\frac{\partial\alpha}{\partial t} = \frac{\partial\alpha}{\partial t} = Du\nabla^2\alpha + a\alpha + b\beta \tag{2.3a}$$

$$\frac{\partial\beta}{\partial t} = Dv\nabla^2\beta + c\alpha + d\beta \tag{2.3b}$$

where $\alpha=u-u_0$, $\beta=v-v_0$, $a=\frac{\partial R^u}{\partial u}$, $b=\frac{\partial R^v}{\partial v}$, $c=\frac{\partial R^v}{\partial u}$, and $d=\frac{\partial R^v}{\partial v}$.

If it is assumed that the solution to equations (2.3) are of the general form $e^{wt}\cos\nu_x x\cos\nu_y y$, then the eigenvalues of the system satisfy

$$[w + D_u\nu^2 - a][w + D_v\nu^2 - d] - cb = 0 \tag{2.4}$$

where $\nu^2=\nu_x^2+\nu_y^2$, and $\nu_x$ and $\nu_y$ are the frequency of perturbation in the x- and y- directions respectively.

Let $$b_1 = \nu^2(D_u+D_v) - (a+d) \tag{2.5}$$

and $$c_1 = D_uD_v\nu^4 - \nu^2(D_va + D_ud) + ad-bc \tag{2.6}$$

then, if can be verified that the system becomes exponentially unstable, i.e., $c_1 < 0$, when $A^2 > k_7k_5^2D/(k_2k_1^2)$, provided $B/D > R_c$ where

$$R_c = \frac{1}{k_3D}\frac{k_2k_1^2A^2+k_7k_5^2D}{k_2k_1^2A^2-k_7k_5^2D}\left[k_4+k_5+k_6\left(\frac{k_1A}{k^5}\right)^2\left(1+\frac{2k_7k_5^2D}{k_2k_1^2A^2+k_7k_5^2D}\right)+D_u\nu^2+\right.$$

$$\left.\frac{D_u}{D_v}\frac{k_2k_1^2A^2+k_7k_5^2D}{k_5^2}+\frac{k_4+k_5}{k_5^2}\frac{k_7k_5^2D+k_2k_1^2A^2}{\nu^2D_v}\right] \tag{2.7}$$

It can easily be shown that $R_c$ attains its minimum value at

$$\nu_*^2 = \frac{1}{D_u}\left[\frac{k_4+k_5^2}{k_5^2}(k_7k_5^2D+k_2k_1^2A^2)\frac{D_u}{D_v}\right]^{1/2} \tag{2.8}$$

In addition, if $A^2 < \frac{k_7k_5^2D}{k_2k_1^2}$, $b_1 > 0$ for all values of (B/D) and the system becomes stable, provided $c_1 > 0$.

Finally, if $A^2 > k_7k_5^2D/(k_2k_1^2)$, $b_1 < 0$ provided $(B/D) > R_c'$, and the system becomes either exponentially or exponential oscillatory unstable depending upon the sign of $b_1^2-4c_1$. Here,

$$R_c' = \frac{1}{k_3D}\frac{k_2k_1^2A^2+k_7k_5^2D}{k_2k_1^2A^2-k_7k_5^2D}\quad k_4+k_5+\nu^2(D_u+D_v)+k_7D+k_2\left(\frac{k_1A}{k_5}\right)^2+$$

$$\left(\frac{k_1 A}{k_5}\right)^2 k_6 \left(1+\frac{2k_7 k_5^2 D}{k_2 k_1^2 A^2 + k_7 k_5^2 D}\right)\Bigg] \qquad (2.9)$$

The above equation is a monotic function of $\nu^2$ and its minimum value occurs at $\nu=0$, provided $A^2 > k_7 k_5^2 D/(k_2 k_1^2)$.

From the above instability conditions we can conclude that

i) The system is unstable for a zero frecuency perturbation ($\nu=0$), provided $(B/D) > R_c'(0)$.

ii) The system is also unstable for nonhomogeneous perturbations ($\nu\neq 0$) provided

$$B/D > R_c'(\nu) \qquad \nu\varepsilon\ (0,\nu^*] \qquad (2.10)$$

$$\text{or} \quad B/D > R_c(\nu) \qquad \nu\varepsilon\ [\nu^*,\infty) \qquad (2.11)$$

where $\nu^*$ is the real positive root obtained by solving $R_c(\nu)=R_c'(\nu)$.

3.- Numerical Solution [5,6]

In order to solve the system of equations (2.1) we employed the following numerical approximation

i) Finite differences.

ii) Galerkin method with the following basis functions;

a) Chapeau,

b) Even Hermite, and

c) Complete Hermite polynomials (Even and Odd).

The definition of the above functions can be found elsewhere [7]. Independently of the method being used for the solution of equations (2.1), the final system of equations, for the continuous time approximation, has the general form

$$\frac{d\alpha}{dt} = A\alpha + B\ F(\alpha,\beta) \qquad (2.12a)$$

$$\frac{d\alpha}{dt} = A\beta + B\ G(\alpha,\beta) \qquad (2.12b)$$

with the associated initial condition $\alpha(0)=\alpha_0$ and $\beta(0)=\beta_0$.

The above system of simultaneous differential equations comprises NM unknowns when using methods (i), (iia), and (iib), and 4(N-1)(M-1) for method (iic), where N and M are the number of mesh points in the x- and y- directions respectively. After discretizing the time, we solved the resulting sytem of nonlinear simultaneous equations by succesive substitution [8], for the nonlinear part, and succesive over-relaxation for the linear part [9].

In order to verify the implementation of the above mentioned methods, we set equal to zero the reaction terms, F and G in equations (2.13) and integrated forward in time the resulting heat diffusion equation using the standard Crank-Nicolson scheme. We compared the results obtained with the various methods, using different initial conditions, against the analytic solution, obtaining the following orders of convergence: a) all time implementations were $O(\Delta t^2)$, b) spatial approximations (i), (iia), and (iib) were $O(\Delta x^2)$ while (iic) was $O(\Delta x^4)$, thus confirming what is reported in the literature [5,6].

## 4.- Numerical Examples

Example 1.- In this first example we set the different parameters involved in the stability criteria in such a way that conditions (2.11b) was satisfied. The value of the parameters were: $D_u/D_v=0.5$, $A=2$, $D=8$, $D_u=1$, $k_1=k_2=k_3=k_4=1$, $k_5=k_6=k_7=k_8=0.5$, $E=8$, and $B/D=6.5$. As for the perturbation frequencies we chose $\nu_x^2=\nu_y^2=2$. With the above parameters the homogeneous steady state is found to be $u_0=4$, and $v_0=12$, as the unique solution to $R^u(u_0,v_0)=R^v(u_0,v_0)=0$. Finally, we fixed the region of interest as $x\ \varepsilon [0,\sqrt{2}\pi]$, and $y\ \varepsilon [0,\sqrt{2}\pi]$, and introduced the following perturbation on $u_0$ and $v_0$ as

$$u(x,y,0) = 4 + 0.2(\cos \nu_x x)(\cos \nu_y y)$$
$$v(x,y,0) = 12.0$$

By observing the time evolution of the system it can be seen that the perturbations introduced to the system grows to a final steady state that exhibited the same frequencies of the perturbation in both directions. The final state reached for the concentration of component u is shown in Figure 1. We investigated further the stability of this final state introducing several small perturbations which died out in time, being this a good indication of its local stability.

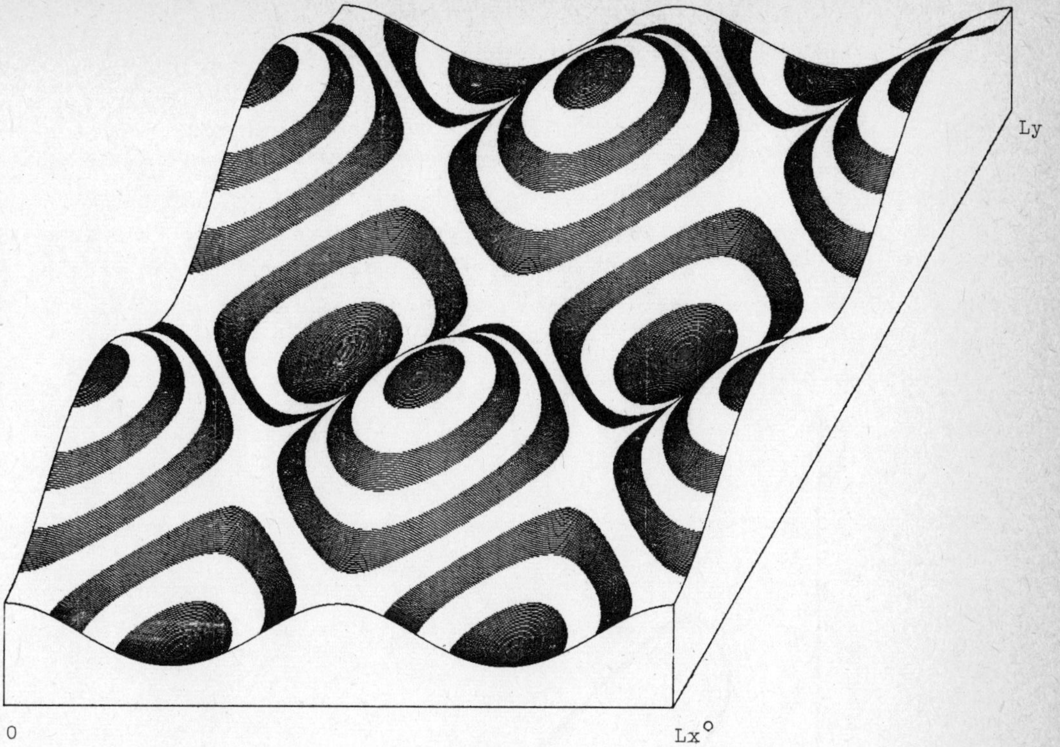


Figure 1.- Two Dimensional Standing Wave (Component U). Example 1.

Example 2.- In this second example, condition (2.11) was also satisfied by choosing the following value for the parameters: $D_u/D_v=1$, $A=2$, $B=8$, $D_u=1$, $k_1=k_2=k_3=k_4=1$, $k_5=k_6=k_7=k_8=0.5$, $E=8$, $B/D=10$, and $\nu_x^2=\nu_y^2= 9/2$. For this set of parameters we found that $u_0=4$, and

$v_0$=17.6. We chose the region to be $x\varepsilon[0,\sqrt{2}\,\pi/3]$, $y\varepsilon[0,\sqrt{2}\,\pi/3]$. Finally, the perturbation introduced to the system were

$$u(x,y,0) = 4 \;+0.05(\cos \nu_x x)(\cos \nu_y y)$$
$$v(x,y,0) = 17.6-0.05(\cos \nu_x x)(\cos \nu_y y)$$

As in the previous example, the system started to move away from the original steady state $(u_0, v_0)$ but instead of retaining the frequencies introduced in the perturbations the system collapsed into an homogeneous limit cycle shown in Figure 2.

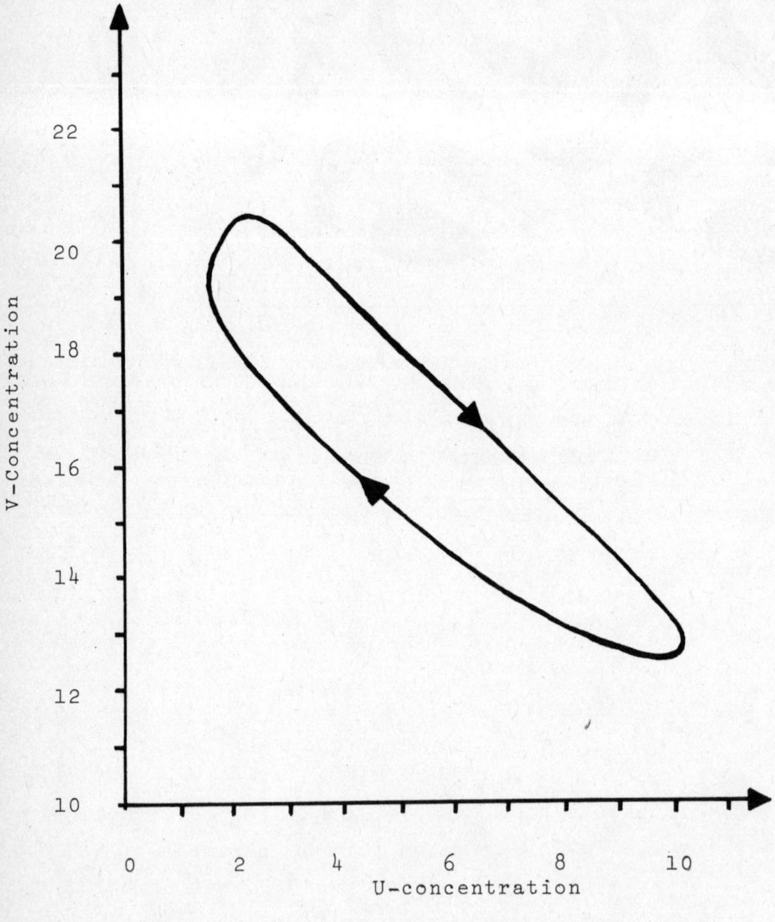


Figure 2.- Limit Cycle for Example 2

Before we present some final comments, it is interesting to remark a mayor difference observed in the performance of the various numerical schemes employed. First of all, for a given partition of the region of interest, schemes (i), (iia) and (iic) produced the same results, but it should be remarked that scheme (iic), namely Galerkin with the complete Hermite polynomials as basis functions, required in order to produce the same results of a time step ten times smaller than that required by the other two schemes. On the other hand, we observed that under no condition the approximation scheme (iib) produced the same results that we obtained by using the other three schemes, that is, the final standing waves for both u and v, were different.

## 5.- Final Comments

In section 2 we derived sufficient conditions for instability of an homogeneous steady state. The stability analysis was carried out by analyzing the characteristic roots of the linearized system of equations governing the time behavior of the system (equations (2.3)). We further verified the conditions derived by the time forward integration of the full nonlinear equations (2.1) thus confirming the validity of the stability criteria previously derived.

For the different numerical schemes employed we found that, for the particular example here analysed, finite differences and Galerkin with chapeau polynomials as basis functions, were superior to Galerkin with both the complete set of Hermite polynomials and the even polynomials as basis functions; several of these results had been previously reported [7].

## References

[1].- Turing, A.M., Phil. Trans. of the Royal Soc. B 237, 37 (1952).

[2].- Prigogine, I., and P. Glansdorff. "Themodynamic Theory of Structure Stability and Fluctuations". Wiley-Interscience (1971).

[3].- Thom, R., "Structural Stability and Morphogenesis; An Outline of a General Theory of Models". W.A. Benjamin, Inc., Reading, Mass (1975).

[4].- Marsden, J.E., and M. McCracken, "The Hopf Bifurcation and its Applications". Springer-Verlag, New York (1976).

[5].- Richtmeyer, R.D., and K.W. Morton, "Difference Methods for Initial Value Problems. Interscience Publishers (1957).

[6].- Douglas, J., and T. Dupont, SIAM J. Num. Anal., 7, 575 (1970).

[7].- Montalvo, A., "The Use of Numerical Methods in Systems with Chemical Reaction and Diffusion". M. Sc. Thesis, Rice University (1972).

[8].- Isaacson, E., and H.B. Keller, "Analysis of Numerical Methods". John Wiley and Sons, Inc., New York (1967).

[9].- Varga, R.S., "Matrix Iterative Analysis". Englewood Cliffs, N. J., Prentice Hall (1962).

# A STUDY OF THE STABILITY OF THE INTERFACE BETWEEN TWO INMISCIBLE VISCOUS FLUIDS

B. CHEN
IIMAS-UNAM
México 20, D.F.

A. NOYOLA
C. Instrumentos-UNAM
México 20, D.F.

1. Introduction. The description of the phenomena of mass transfer across the interface between two inmiscible fluids is very important in the understanding of separation processes, such as evaporation, drying, extraction, etc.

Several researches (Laird (1954), Yih (1955), Hanrotty and Engen (1957), Arriola and Noyola (1975), Schlichting (1968), Shuler (1976) have studied the problem. It has been necessary to do some approximations such as parallel flow. In this work we consider laminar flow, and introduce a small perturbation to the velocities of parallel movement. The behavior of the interface will depend on the Reynolds number of each fluid and on the wavelength of the perturbation. This paper is the first step in determining the region of stability.

2. Formulation of the problem. We will study the linearized stability of two inmiscible, incompressible, viscous fluids between two parallel walls (Poiseuille flow). See figure 1.

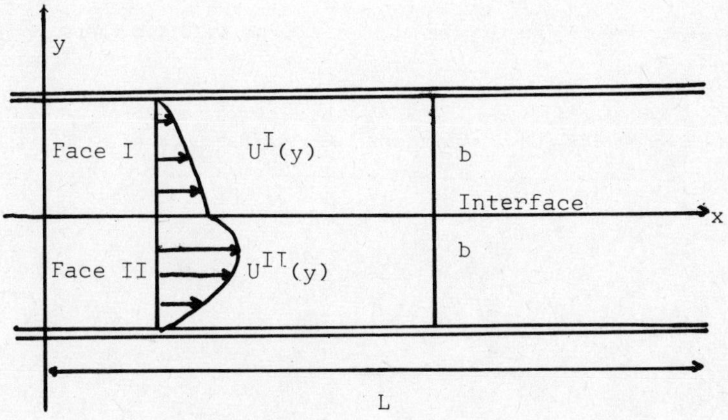


Fig. 1.- Velocity profiles of unperturbed systems.

For constant density, $\rho$, each fluid satisfies the following equations of motion, (Schlichting (1968), Bird, Steward and Lightfoot (1960)),

$$\frac{\partial u}{\partial x} + \frac{\partial v}{\partial y} = 0 \tag{1}$$

$$\rho\left(\frac{\partial u}{\partial t} + u\frac{\partial u}{\partial x} + v\frac{\partial u}{\partial y}\right) = -\frac{\partial p}{\partial x} + \mu\left(\frac{\partial^2 u}{\partial^2 x} + \frac{\partial^2 u}{\partial^2 y}\right) \tag{2}$$

$$\rho\left(\frac{\partial v}{\partial t} + u\frac{\partial v}{\partial x} + v\frac{\partial v}{\partial y}\right) = -\frac{\partial p}{\partial y} + \mu\left(\frac{\partial^2 v}{\partial^2 x} + \frac{\partial^2 v}{\partial^2 y}\right) + \rho g \tag{3}$$

Here u and v are the cartesian components of the velocity (there is no z dependence), $\rho$ is the presure, $\mu$ is the dynamic viscosity and $g$ the acceleration of gravity.

The boundary conditions are:

1) at the walls u=v=0
2) in the interface:
   a) The normal stress has a discontinuity proportional to the curvature of the interface.
   b) The tangential stress is continuous.
   c) The tangential velocity is continuous.
   d) There is no fluid transfer across the interface.

We will decompose the motion into a mean flow (whose stability we want) and into a small disturbance superimposed on it. The x and y components of the mean velocity are denoted by U and V, respectively, and the pressure by P. The corresponding qualities of the perturbation are $\hat{u}$, $\hat{v}$ and $\hat{p}$, so that the whole velocity is

$$u = U + \hat{u} \quad , \quad v = V + \hat{v}$$

and the pressure is

$$p = P + \hat{p}$$

When there is need to specify one or the other fluid, we will use the superindices I or II.

To solve for the mean motion (no perturbation) we consider a parallel flow, that is

$$V^i = 0 \tag{4}$$

$$U^i = U^i(y) \qquad i = I, II \tag{5}$$

so that the Navier Stokes equations are reduced to

$$\mu^i \frac{d^2 U^i}{d y^2} = \frac{\partial P^i}{\partial x} \tag{6}$$

$$\frac{\partial P^i}{\partial y} = \rho^i g \tag{7}$$

whose solution is

$$P^i = \rho^i g y + \pi^i(x) \tag{8}$$

$$U^i = \frac{1}{2\mu^i} \frac{\partial P}{\partial x} y^2 + C_1 y + C_2 \tag{9}$$

With $C_1$, $C_2$ constants and $\frac{\partial P}{\partial x} = \frac{\partial P^I}{\partial x} = \frac{\partial P^{II}}{\partial x}$ constant from the conditions at the interface.

Or normalizing

$$U^I = U^\circ(1-\lambda_1^I \chi - \lambda_2^I \chi) \qquad 0 \leqslant \chi \leqslant 1 \tag{10}$$

$$U^{II} = U^\circ(1-\lambda_1^{II} \chi - \lambda_2^{II} \chi^2) \qquad -1 \leqslant \chi \leqslant 0 \tag{11}$$

where

$$U^\circ = \frac{\partial P}{\partial x} \frac{b^2}{\mu^I + \mu^{II}} \tag{12}$$

$$\lambda_1^i = \frac{\mu^I - \mu^{II}}{2\,\mu^i} \tag{13}$$

$$\lambda_2^i = \frac{\mu^I = \mu^{II}}{2\,\mu^{II}} \tag{14}$$

$$\chi = \frac{y}{b} \tag{15}$$

3. Stability analysis. For the linearized stability of the mean flow, the perturbation is considered small.

Substituting

$$u^i = U^i + \hat{u}^i, \quad v^i = \hat{v}^i, \quad p^i = P^i + \hat{p}^i \quad , \quad i = I, II \tag{16}$$

into the Navier-Stokes equations, neglecting terms of second order and lower, and using the results for the mean flow we get

$$\frac{\partial \hat{u}^i}{\partial t} + U^i \frac{\partial \hat{u}^i}{\partial x} + \hat{v}^i \frac{dU^i}{dy} + \frac{1}{\rho^i} \frac{\partial \hat{p}^i}{\partial x} = \nu^i \nabla^2 \hat{u}^i \tag{17}$$

$$\frac{\partial \hat{v}^i}{\partial t} + U^i \frac{\partial \hat{v}^i}{\partial x} + \frac{1}{\rho^i} \frac{\partial \hat{p}^i}{\partial y} = \nu^i \nabla^2 \hat{v}^i \tag{18}$$

$$\frac{\partial \hat{u}^i}{\partial x} + \frac{\partial \hat{v}^i}{\partial y} = 0 \tag{19}$$

where $\nu^i = \mu^i/\rho^i \quad , \qquad i = I, II \qquad (20)$

Taking the y derivative of (17) and subtracting the x derivative of (18), and using the following change of variables to normalize:

$$\tau = \frac{U^\circ t}{b} \quad , \quad \chi = \frac{y}{b} \quad , \quad \xi = \frac{x}{b}$$

$$u^i = \frac{\hat{u}^i}{U^\circ} \quad , \quad v^i = \frac{\hat{v}^i}{U^\circ} \quad , \quad U^i = \frac{U^i}{U^\circ} \quad , \quad R^i = \frac{U^\circ b}{\nu i} \tag{21}$$

we obtain

$$\frac{\partial}{\partial\tau}\left(\frac{\partial}{\partial\xi}\tilde{v}^i - \frac{\partial\tilde{u}^i}{\partial\chi}\right) + \tilde{U}^i\frac{\partial^2\tilde{v}^i}{\partial\xi^2} - \tilde{v}^i\frac{\partial^2\tilde{U}^i}{\partial\chi^2} - \tilde{U}^i\frac{\partial^2\tilde{u}^i}{\partial\chi\partial\xi} =$$

$$= \frac{\partial}{R^i}\left[\frac{\partial^3\tilde{v}^i}{\partial\chi^2\partial\xi} + \frac{\partial^3\tilde{v}^i}{\partial\xi^3} - \frac{\partial^3\tilde{u}^i}{\partial\chi^3} - \frac{\partial^3\tilde{u}^i}{\partial\xi^2\partial\chi}\right] \tag{22}$$

Introduce a stream function $\psi^j$, j=I, II, for each of the fluids

$$\psi^j = \varphi^j(\chi)\ ^{i(\alpha\xi - \beta\tau)} \tag{23}$$

so that

$$\tilde{v}^j = -\frac{\partial\psi^j}{\partial\xi} \quad , \quad \tilde{u}^j = \frac{\partial\psi^j}{\partial\chi} \tag{24}$$

This stream function represents a single oscillation of the disturbance. We assume that it is possible to expand an arbitrary perturbation in a Fourier series. Here $\alpha$ is the real wavenumber of the disturbance and $\beta$ is $\alpha$ times the complex wavespeed. If the imaginary part of $\beta$ is negative, the disturbances are damped, and the flow is stable, otherwise, if it is positive the flow is unstable. $\varphi(\chi)$ is the complex amplitude is assumed to depend only on $\chi$ since the mean flow depends only on $\chi$ (Schlichting (1968)).

Subtituting (23) and (24) into (22) we get

$$\varphi^j(\chi) + \varphi^j(\chi)[-(U^j-C)i\,\alpha\,R^j - 2\alpha^2]$$

$$+\varphi^j(\alpha^4 + (U^j-C)i\,\alpha^3\,R^j + U^j\,i\,\alpha\,R) = 0 \tag{25}$$

with

$$C = \frac{\beta}{\alpha} = C_r + i\,C_i \tag{25}$$

(25) is a system of two Orr-Sommerfeld equations.

The boundary conditions for (25) are obtained linearizing the boundary conditions for the physical problem, Wehausen and Laitone, (1960). The conditions at the interface are evaluated at $\chi$=0.

$$\begin{aligned}
\varphi^I(1) &= \varphi^{I'}(1) = 0 && \text{(zero velocity at topwall)}\\
\varphi^{II}(-1) &= \varphi^{II''}(-1) = 0 && \text{(zero velocity at bottomwall)}\\
\varphi^I(0) &= \varphi^{II}(0) = 0 && \text{(normal velocities at interface)}\\
\varphi^{I'}(0) &= \varphi^{II'}(0) = 0 && \text{(tangential velocities at interface)}\\
\mu^I\varphi^{II''}(0) &= \mu^{II}\varphi^{I''}(0) && \text{(tangential stresses at interface)}
\end{aligned} \tag{27}$$

We have a system of 2 fourth order complex differential equations with 8 complex boundary conditions.

4. Numerical Solution. The stability of a laminar boundary layer of an incompressible fluid, is given by the solution of an Orr-Sommerfeld equation. Among the many method proposed for finding its eigenvalues and eigenvectors, it is worthwile to mention finite differences (Osborne (1967), Jordinson (1970), Lentini and Pereyra (1977)), multiple shooting (England (1976)), Chebyshev polynomials (Orszag (1971)), ortonormalization, pseudo-orthonormalization (Mack (1976)), Monkewitz (1978)). To our knowledge nothing has been done for a system of Orr-Sommerfeld equations. We tested the methods of finite differences (Lentini and Pereyra (1977)) and multiple shooting. Both methods required about the same computing time and memory when a sufficiently good initial guess was given. However England's program uses Marquard's algorithm when Newton doesn't converge and Marquard is very slow. Lentini's and Pereyra's program (PASVA 3) has a Newton scheme with step control that converges for a reasonable first guess. One of the main problems was the lack of an initial guess. So we picked PASVA 3 for this reason.
To have a two point boundary value problem, it is necessary to change the independent variable $\chi$ in region II to $-\chi$ so that the problem be defined in $\chi \in [0,1]$. Next the eigenvalues $C_R$ and $C_I$ are considered as unknowns, and two additional equations are introduced

$$C_R' = 0$$
$$C_I' = 0 \qquad (28)$$

together with two additional boundary conditions to normalize the solution.

Next we write the system of equations as a system of 18 first order real differential equations. The range of convergence of Newton's iteration is very small. Step control increases it somewhat, but a fairly good first guess is still necessary. To obtain one we solve for a nonsymmetric mode of Poiseuille flow for one fluid, for which solutions are known, and then use a continuation procedure to obtain solutions for two fluids.

5. Numerical Results. For one fluid we were able to calculate solutions for Reynolds number of about 3000 with no difficulty. For two fluids, octane on top of water, we have only calculated the main mode for fairly low Reynolds numbers. See figure 2. This solution is stable. We are in the process of increasing the Reynolds numbers until we find the point of neutral stability. An open question is how to calculate all the modes of the solution.

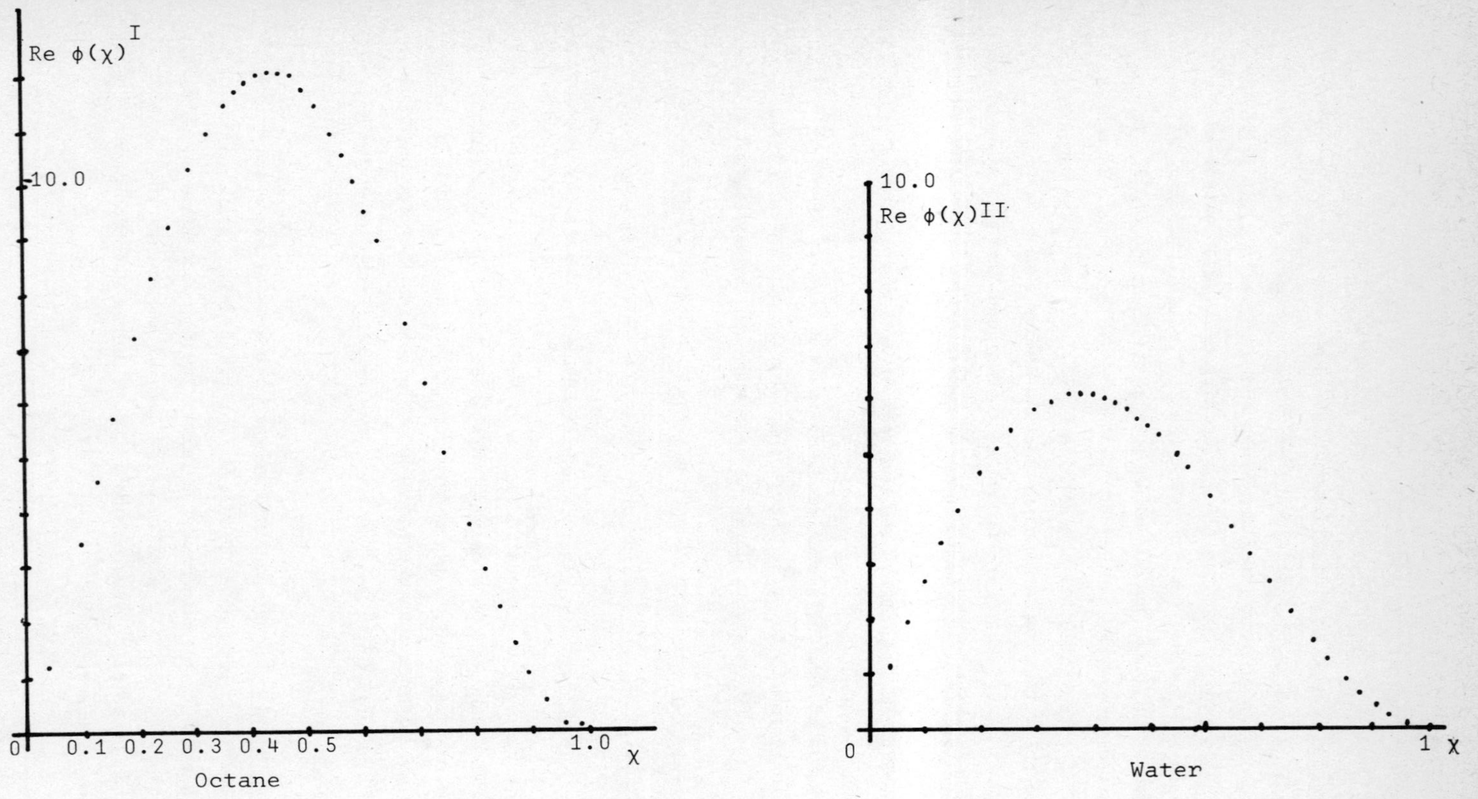


Fig. 2.- Variation of Re $\phi(\chi)$ with $\chi$ when $R^{I}=4.152$, $R^{II}=2.239$, $C_R=0.7293$, $C_I=-6.7986$, $U_0=1$ cm/seg $b=2.25$ cm., $\alpha=1.0$, $\mu^{I}=1.003$ gr/cm seg, $\mu^{II}=0.380$ gr/ cm seg, $\rho^{I}=0.998$ gr/cm$^3$, $\rho^{II}=0.702$ gr/cm$^3$

## References

1. Arriola, A.T. and A.I. Noyola, Transferencia de masa a través de una interfase formada por dos fluídos inmiscibles en condiciones hidrodinámicas, Rev. del I.M.P. Vol. VII, No. 3, 1975.

2. Bird, B.R., W.E. Steward and E.N. Lightfoot, Transport phenomena, John Wiley & Sons, New York, 1960.

3. Davey, A., On the numerical solution of difficult boundary-value problems.J. Comp. Phys. 35, 1980.

4. England, R., A program for the solution of boundary value problems for systems of ordinary differential equations. Report CLM-PDN 3/73 Culham Laboratory, Abingdon Oxfordshire, 1976.

5. Gersting, J.M., Numerical methods for eigensystems: the Orr-Sommerfeld problem as an initial value problem, Comp. and Math. Appls. 6, 1980.

6. Hanrctty, T.S. and J.M. Engen, The adjacent flow of a turbulent and a laminar layer, A.I.Ch.E.J. 3, 1957.

7. Jordinson, R., The flat-plate boundary layer. Part I. Numerical integration of the Orr-Sommerfeld equation, J. Fluid Mech. 50, 1970.

8. Laird, A.D.K., Annular gas liquid flow in tubes, Trans. ASME 76, 1954.

9. Lentini, M. and Pereyra, V., An adaptive finite difference solver for nonlinear two-point boundary problems with mild boundary layers, SIAM J. Numer. Anal. 14, 1977.

10. Mack, L.M., A numerical study of the temporal eigenvalue spectrum of the Blasius boundary layer, J. Fluid Mech. 73, 1976.

11. Monkewitz, M.A., Analytic pseudoorthogonalization melthods for linear two-point boundary value problems illustrated by the Orr-Sommerfeld equation, ZAMP 29, 1978.

12. Orszag, S.A., Accurate solution of the Orr-Sommerfeld stability equation, J. Fluid Mech. 50, 1971.

13. Osborne, M.R., Numerical methods for hydrodynamic stability problems, SIAM J. Appl. Math. 43, 1970.

14. Schlichting, H., Boundary-Layer Theory, McGraw-Hill, New York, 1968.

15. Shuler, P.J. and W.B. Krautz, The equivalence of the spatial and temporal formulation for the linear stability of falling film flow, A.I.Ch.E.J. 22, 1976.

16. Stijus, Th.L. and Van de Vooren, An accurate method for solving the Orr-Sommerfeld equation, J.Eng.Math. 14, 1980.

17. Wehausen, J.V. and E.V. Laitone, Surface Waves, in Handbuck der Physick, IX, 1960.

18. Yih, C.S., Stability of two-dimensional parallel flows for three dimensional disturbances. Quart. Appl. Math. 12, 1955.

# SOLVING LARGE NONLINEAR SYSTEMS OF EQUATIONS ARISING IN MECHANICS

JORGE NOCEDAL

IIMAS - UNAM
Apdo. Postal 20-726
México 20, D.F.
MEXICO

## 1. INTRODUCTION

The purpose of this paper is to study the numerical solution of nonlinear systems of equations with the following properties: they are large and sparse, mildly nonlinear and have a relatively simple sparsity structure. We adapt some algorithms developed for small dense problems, such as quasi-Newton and Newton-like methods, to this class of problems. First, their storage requirements are reduced, and second, the computation of the displacements is performed inaccurately in order to reduce the amount of work involved.

The problems we consider arise in nonlinear structural analysis and can be cast either as optimization problems or as nonlinear systems or equations. In the latter case the Jacobian (or stiffness) matrix is symmetric and usually positive definite. Therefore for computational purposes both approaches are basically the same. For instance, the BFGS method for minimization can be applied to the solution of nonlinear equations with a change only in the line search procedure (if at all). For definiteness let us write the problem as

$$\min \quad f(u). \tag{1}$$

Here u could represent the displacements at the nodal points within a structure, and the system of equations $\nabla f(u)=0$ defines the equilibrium configuration.

The most suitable algorithms for large scale optimization are: a) the conjugate gradient (CG) method; b) sparse quasi-Newton methods; c) variable storage methods and d) Newton-like methods. We will describe them in section (4), except for the sparse quasi-Newton methods. They are not considered here because we feel that they have little to offer for our type of problems. The goal of this paper is to point out some basic characteristics of the problems, that are relevant to the design of numerical methods.

## 2. A SIMPLE EXAMPLE

Let us consider the following problem. It represents a plate that is clamped on one edge and is loaded on the opposite side. It is given by

$$\min f(u) = \sum_{i=1}^{N} \sum_{j=1}^{N} [\tfrac{1}{2}(u_{i,j}-u_{i,j-1})^2 + \tfrac{1}{2}(u_{i,j}-u_{i-1,j})^2$$

$$+ \tfrac{1}{2} N^2 (u_{i,j}-u_{i,j-1})^4 + \tfrac{1}{2} N^2 (u_{i,j}-u_{i-1,j})^4]$$

$$- \beta \sum_{j=0}^{N} c_j u_{N,j} \qquad (2)$$

with $u_{0,j}=0$, $j=0,\ldots,N$. The distribution of the load is given by $c_j$ and the parameter $\beta$ determines the weight of the load. The Hessian matrix at $u=0$ has the form

$\nabla^2 F(0) =$

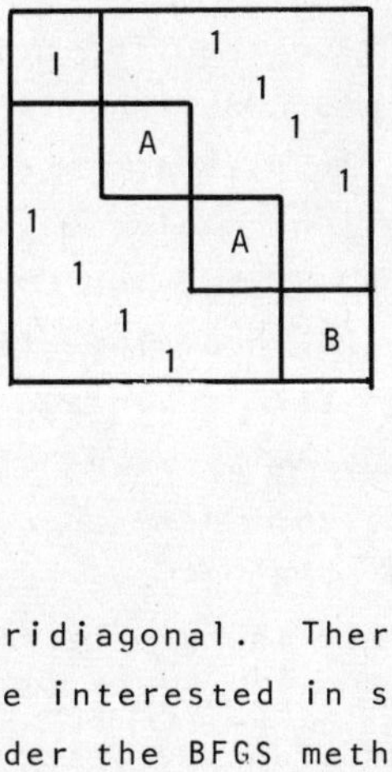


where A and B are tridiagonal. Therefore $\nabla^2 f(0)$ is an $N^2 x N^2$ pentadiagonal matrix. We are interested in solving the problem for N=20 and N=50. Let us consider the BFGS method:

$$x_{k+1} = x_k - H_k \nabla f(x_k) \qquad (3)$$

$$H_{k+1} = (I - \frac{s_k y_k^T}{s_k^T y_k}) H_k (I - \frac{y_k s_k^T}{s_k^T y_k}) + \frac{s_k s_k^T}{s_k^T y_k} . \qquad (4)$$

Here $s_k := x_{k+1} - x_k$, $y_k := \nabla f(x_{k+1}) - \nabla f(x_k)$ and we write now $x_k$ instead of $u_k$ to be consistent with the optimization literature. The initial point $x_0$ can be chosen as the undeflected plate, i.e., $x_0=0$. There are several possibilities for the initial quasi-Newton matrix, for example, $H_o=I$ or

$H_o$=[main tridiagonal of $\nabla^2 f(0)]^{-1}$. To try them experimentally we set $\beta$=.1, N=20 and stop the iteration when $\|\nabla f\|_2 \leqslant 10^{-4}$. We note that the weight $\beta$=.1 produces a deflection at the tip, of the same magnitude as the side of the plate (the distance between the clamped and the loaded edges). For $H_o$=I we need 288 iterations; for $H_o$= [main tridiagonal of $\nabla^2 f(0)]^{-1}$ we still need 210 iterations. Finally, we use $H_o=\nabla^2 f(0)^{-2}$ and to our surprise we find the solution in 4 iterations. This seems to be the definite answer except for the fact that the LU decomposition of $\nabla^2 f(0)$ has a full band and the solution of the linear system

$$\nabla^2 f(0)\ \Delta_k = -\nabla f(X_k) \qquad (5)$$

is rather expensive for large N. For N=50 it is prohibitively costly. Therefore we may opt for the following: use the good initial matrix but solve the system (5) only approximately. This approach has been rather successful in various applications; see for example Concus, Golub and O'Leary [2]. Now we have to answer two questions. What method would be appropriate for solving the linear system? How accurately should we solve it? The second question is not easy and will be discussed later. Regarding the first, there are many iterative methods for linear systems and the right choice depends on the nature of the problem (see Young [11]). In our case we could choose between SSOR [11], conjugate gradients [5] and incomplete factorization [7]. Let us use SSOR by blocks (BSSOR) to take advantage of the block structure of $\nabla^2 f(0)$. If we solve the problem by means of the BFGS method with limited storage, (see next section) where the solution of (5) is approximated by doing only 2 sweeps of BSSOR, we require 49 iterations and the total amount of work performed is much less than in the previous runs. It remains to be seen whether this algorithm can be improved and if it is successful for other problems. Before doing this let us take a closer look and make an operation count for the different stages of the algorithm. Introducing a work unit u to normalize the cost of a function and gradient evaluation we have, roughly:

| STAGE | COST |
|---|---|
| 1. Function and gradient evaluation | 1 u |
| 2. One BSSOR sweep | 1 u |
| 3. BFGS iteration (storing 5 corrections) | 2 u |
| 4. Conjugate gradient iteration | 0.4 u |
| 5. Hessian evaluation | 0.3 u |

In testing algorithms for small problems it is frequently assumed that the cost of evaluating the function and gradient dominate. Many codes have been tuned up so as to do as few function evaluations as possible. Here the situation is quite different and we should revise the implementation of the algorithm, in particular the line search procedure, and consider the total amount of work; not only the function and gradient evaluations. We also see that for this problem to compute the Hessian is inexpensive and, probably, one should prefer Newton-like algorithms over quasi-Newton methods. In fact one finds experimentally that Newton-BSSOR is more efficient than BFGS-BSSOR. For many problems the evaluation of the Hessian will be much more costly than in this example. In that case it may prove wise to compute the Hessian only at certain iterations and to fix it during the rest. This will be discussed in the next section.

## 3. APPROXIMATIONS

Let us suppose that all the algorithms to be studied are of the form

$$x_{k+1} = x_k + \alpha_k \Delta_k \tag{6}$$

where $\alpha_k$ is a steplength and $\Delta_k$ is an approximate solution of the linear system

$$B_k \Delta_k = -\nabla f(x_k) \tag{7}$$

Here $B_k$ may be the Hessian or a quasi-Newton matrix. The point now is that if (7) is to be solved inaccurately then there is no reason for requiring $B_k$ to be exact either. Ortega and Rheinboldt [9] give some results in this respect. If $\|B_k - \nabla^2 f(x_k)\| \to 0$ then the iteration (6)-(7), with (7) solved exactly, is superlinearly convergent under some standard assumptions. If for all k $\|B_k - \nabla^2 f(x_k)\| \leqslant \alpha \|\nabla f_k\|$, for some constant $\alpha$, then the convergence is quadratic. These conditions may lead to practical implementations in the case where the difference between the Hessian and the approximating matrix is easy to measure. This is the case in some problems and the main guide will be a physical consideration. For the problems of Section 5, for example, the elastic-plastic properties of the components are described by the following graph

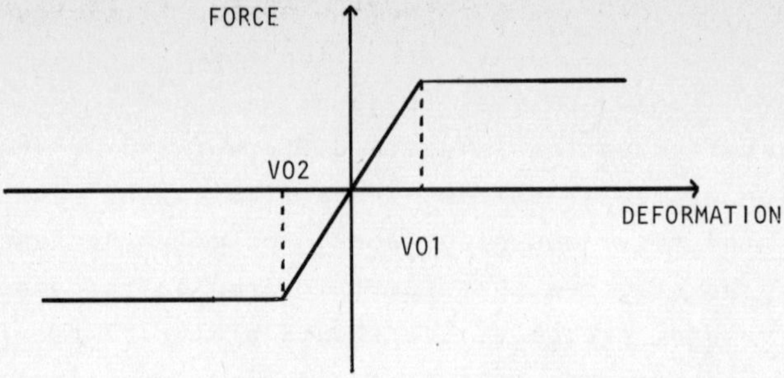


where V01 and V02 are parameters that specify various materials ( cables and rods in this case). The space of deformations can be divided into three regions: $(-\infty, V02)$, $[V02, V01]$, $(V01, \infty)$. One can see that the stiffness matrix changes only if the deformation passes from one region into another. This can be determined with very little expense.

It happens also in some applications that only one part of the Hessian changes. This would happen, for example, when only one part of the structure is working at that stage. A reasonable algorithm would only reevaluate this part of the Hessian. One application of these ideas leads to the following algorithm. Choose a constant $\alpha$ and evaluate the Hessian. Apply modified Newton's method (or BFGS, or CG) until the estimated difference between old and new Hessian is greater than $\alpha \| \nabla f(X_k) \|$. Then recompute all or part of the Hessian, depending on what has occurred in the intermediate iterations. Numerical experiments have been performed using such an algorithm. The results will be reported in a separate paper, together with a detailed description of the problems.

Let us now consider how accurate should the solution of (7) be. Dembo, Eisenstat and Steihaug [4], have studied this for Newton's method. One has for example, that if the residual $r_k := B_k \Delta_k + \nabla f(x_k)$ satisfies

$$\| r_k \| \leqslant c \, \| \nabla f(x_k) \|^2 , \qquad (8)$$

and if $B_k$ is the Hessian, then the rate of convergence is quadratic. A difficulty here is that the constant c in (8) has no clear physical meaning. In general one would have to chose it arbitrarily. In our numerical experiments we found that the choice of c changed the efficiency of the algorithms considerably. In the case of BSSOR we opted for doing a fixed number of sweeps throughout. For the Newton-SOR method Ortega and Rheinboldt [9] give a rate of convergence result that depends directly on the number of SOR sweeps. Let $\rho(\cdot)$ denote the spectral radius and let H be the operator associated with SOR (see [9]), then under standard continuity and nonsingularity assumptions, and the additional assumption that at the solution $x^*$ $\rho[H(x^*)] < 1$, the rate of convergence is

$$R_1(x_k) \leqslant \rho\,[H(x^*)]^{m'} \quad , \qquad (9)$$

where

$$m' = \lim_{k\to\infty} \inf \; m_k \; ,$$

and $m_k$ is the number of SOR iterations made at the k-th step. Both (8) and (9) indicate the importance of solving the linear system accurately towards the end of the process. However, one of the characteristics of the class of problems in structural analysis under consideration is that a low accuracy in the answer is acceptable; e.g. $10^{-2}$ or $10^{-4}$. Superlinear and quadratic convergence could be undistinguishable in this case and one could be wasting effort trying to obtain a faster algorithm without succeeding in it. The question of how to chose practically the accuracy in the solution of the linear system is not simple and perhaps has to be answered for each problem by trial and error. However, as discussed in the next section it is not crucial; of greater importance is the choice of the iteration matrix $B_k$.

## 4. NUMERICAL METHODS

To keep low the storage requirements of the BFGS method one can store the corrections separately, instead of overwriting the matrix. Note that each update of the matrix (4) is defined by two n-vectors, $s_k$ and $y_k$. When the product of $H_k$ and $\nabla f(x_k)$ is needed one can perform it efficiently using a scheme of Matthies and Strang [6]. After a certain number of corrections have been stored (e.g.,5) we will not want to increase the storage further. A possibility is to bring in a new correction and discard the oldest one. An algorithm for doing this is given by Nocedal [8] and we will call it the Limited Storage BFGS. It keeps always a fixed number of corrections, updating them at every step by deleting the oldest one and introducing a new one. Our implementation uses a line search procedure consisting of safeguarded cubic interpolation (see Shanno and Phua [10]). The preconditioned conjugate gradient method [1], with the same line search routine, has also been tried experimentally and found competitive with the Limited Storage BFGS.

We present now some results to show the effect of the accuracy of the solution of the linear system in the overall efficiency of the algorithms. We tried the Limited Storage BFGS and the CG on problem (2) with N=20. The linear system was solved inaccurately by doing a fixed number of BSSOR iterations. The relative cost units were taken as follows: (a) 1 function and gradient evaluation=1u; (b) 1 BSSOR iteration=1u; (c) BFGS iteration, keeping 5 corrections=2u; (d) CG iteration=0.4u. The relaxation parameter for the BSSOR iteration was taken as 1.5 and the condition to stop the algorithms was $\|\nabla f(x_k)\| \leq 10^{-4}$. The initial matrix for the BFGS and the preconditioner for the CG was $B_o=\nabla^2 f(0)$. For comparison we include results for Newton's method, with and without a line search.

| | BFGS | | | | CG | | | |
|---|---|---|---|---|---|---|---|---|
| No. of SSOR | $\beta=5\times10^{-3}$ | | $\beta=.1$ | | $\beta=5\times10^{-3}$ | | $\beta=.1$ | |
| SWEEPS | ITER/FUN | COST | ITER/FUN | COST | ITER/FUN | COST | ITER/FUN | COST |
| 1 | 21/25 | 88 | 70/73 | 283 | 23/49 | 82 | 46/49 | 159 |
| 2 | 10/15 | 54 | 49/51 | 247 | 16/35 | 74 | 36/73 | 159 |
| 4 | 9/13 | 67 | 22/24 | 156 | 10/23 | 67 | 25/51 | 161 |
| 8 | 8/15 | 95 | 17/20 | 190 | 6/15 | 66 | 18/37 | 188 |
| Newton Line Search | 2/7 | | 4/9 | | | | | |
| Newton No Line Search | 10/11 | | 8/9 | | | | | |

For N=50 the results are the following:

$\beta=5\times10^{-3}$

| No. of SSOR | BFGS | | CG | |
|---|---|---|---|---|
| SWEEPS | ITER/FUN | COST | ITER/FUN | COST |
| 1 | 45/49 | 184 | 61/127 | 212 |
| 2 | 28/30 | 142 | 31/66 | 140 |
| 4 | - - | - | 28/59 | 182 |

Note that doing 2 BSSOR sweeps per iteration is always better than just one. For this particular problem 2 or 4 seem optimal.

We tried several values for the relaxation parameter in the SSOR iteration and found that the methods are not very sensitive to it. This had been noted previously by Concus, Golub and O'Leary [2]. We also explored, using various problems, the effect of the steplength found by the one dimensional search. The routine of Shanno and Phua [10] uses a rather slack condition for accepting the step. In our experience it does not seem better, in general, to make a more accurate line search. At any rate the efficiency of the methods does not depend very strongly on these issues. It seems that the choice of the iteration matrix and the way of solving the linear system are by far the most important factors. For problem (2) one can show that if $x_o=0$ and $B_o=I$ then BFGS or CG will need at least N iterations to find the solution, regardless of the value of β. In practice we find that for N=20 and β=.1 BFGS needs 288 iterations. This is extremely inefficient, and there is something more disturbing. As the value of β is decreased the problem becomes simpler. However, BFGS will always need more than N iterations because the information collected at each step is deficient. The quasi-Newton matrix is built very slowly and one needs the full N-steps to obtain a reasonable iteration matrix. This leads us to think that for large sparse problems one should select the information along which the information is collected, as is the case in the Curtis, Powell, Reid [3] strategy for finite differences. The selection would depend on the sparsity structure of the problem. It is an open question if a robust quasi-Newton method following these ideas can be devised.

## 5. A CLASS OF TEST PROBLEMS

In this section we will describe a class of engineering structures that generate large nonlinear optimization problems. We have found them rather useful because the degree on nonlinearity can be changed in various ways and the resulting linear systems can be made to have a variable density. The structures are composed of rods and cables and we assume that the cables can only be stretched (and not compressed) and that the rods can only be compressed (and not stretched). Here is a typical element

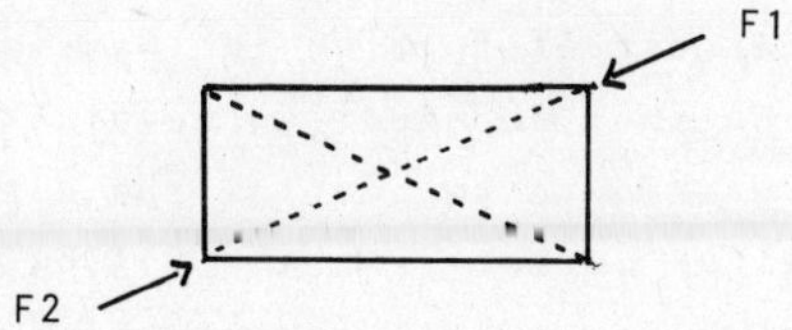

The dotted lines represent cables and the solid lines rods that are hinged together. After applying the forces $F_1$ and $F_2$ we obtain the following configuration

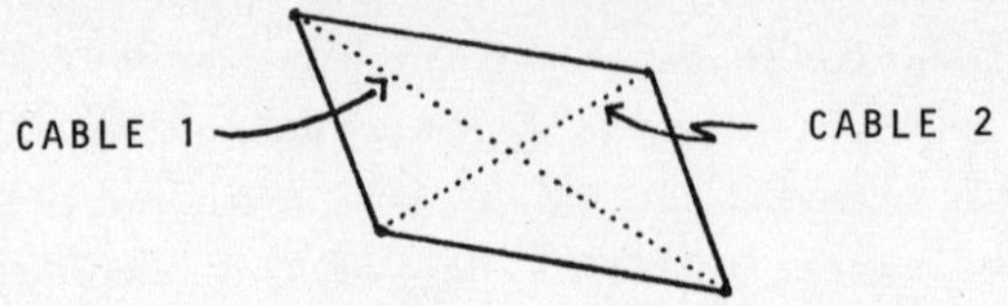


Cable 1 is not working and cable 2 is stretched. Complex structures can be constructed using these elements or similar triangular elements.

Several examples of these structures and numerical experiments using quasi-Newton and Newton type methods will be reported in a separate paper. Here we will only discuss how the ideas of the previous sections can be applied in this case. Take for example structure A.

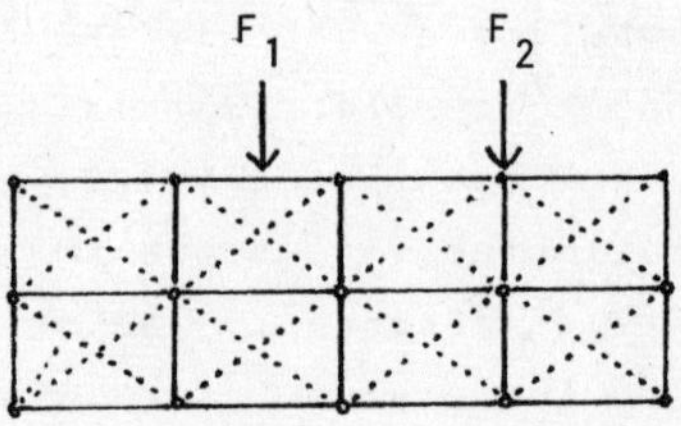


During the various faces of the optimization process different parts of the structure will start working, until the final configuration is obtained. The result is that the Hessian matrix will be changing by blocks and that we can know when the Hessian has varied by noting if an element stopped or restarted working. Another interesting feature of these problems is that the Hessian is sometimes singular so that Newton-like methods cannot be applied directly. The Hessian is also banded and its size and bandwidth depend on the number of elements and the number of rows of the structure. How should we choose the iteration matrix in these problems? For the case where the Hessian is nonsingular a modified Newton method, with the linear system solved by SSOR, is appropriate. The cost of forming the Hessian matrix is not great and only the part that has changed is reevaluated. Is is useful to include a line search procedure to control the displacements. For singular problems a BFGS-SSOR method with the following initial matrix has proved to be efficient. Remove one (or more) of the outer bands of the Hessian until a nonsingular (diagonally dominant) matrix is obtained. Use this as the initial matrix.

Finally, we will note that the usual form of constructing quasi-Newton matrices, that is, by imposing the secant equation, does not take advantage of the block structure of the Hessian. Could one devise secant conditions that apply by blocks and generate well behaved quasi-Newton matrices? This could be particularly useful in problems of this type.

ACKNOWLEDGEMENTS

The author is grateful to Gilbert Strang for may discussions and for suggesting the problem os Section 2, and to Jan Holnicki-Szulc for providing the structures of Section 5.

## REFERENCES

[1] Axelsson, O.; On preconditioning and convergence acceleration in sparse matrix problems. Report 74-10, Data Handling Division, CERN, Geneva (1974).

[2] Concus, P., Golub, G., O'Leary, D.; Numerical solution of nonlinear elliptic partial differential equations by a generalized conjugate gradient method, Computing 19, 321-339 (1978).

[3] Curtis, A., Powell, M., Reid, J.; On the estimation of sparse Jacobian matrices, J. Inst. Math. Appl. Vol. 13, 117-119, (1974).

[4] Dembo, R., Eisenstat, S., Steihaug, T.; Inexact Newton methods, Tech. Report Series B: No. 47, School of Organization and Management, Yale University (1980).

[5] Hesteness, M., Stiefel, E.; Methods of conjugate gradients for solving linear systems. J. Res. Nat. Bur. Stand. 49, 409-436 (1952).

[6] Matthies, H., Stang, G.; The solution of nonlinear finite element equations, Inter. J. of Num. Meth. in Eng. Vol. 14, (1979).

[7] Meijerink, J., van der Vorst, H.; An iterative solution method for linear systems of which the coefficient matrix is a symmetric M-matrix. Math. Comp. 31, 148-162 (1977).

[8] Nocedal, J.; Updating quasi-Newton matrices with limited storage, Math. Comp. Vol. 35, 773-782 (1980).

[9] Ortega, J., Rheinboldt, W.; Iterative solution of nonlinear equations in several variables, Academic Press (1970).

[10] Shanno, D., Phua, K.; A variable method subroutine for unconstrained nonlinear minimization, MIS. Tech. Rep. No. 28, University of Arizona (1978).

[11] Young, D.; Iterative solution of large linear systems, Academic Press (1971).

# SMOOTH MONOTONE SPLINE INTERPOLATION

G. Pagallo and V. Pereyra

Escuela de Computación
Facultad de Ciencias
Universidad Central de
Venezuela , Caracas

## 1. INTRODUCTION

In the course of designing an automatic mesh refinement procedure for solving two-point boundary value problems for ordinary differential equations, the authors [3] came accross the need to produce smooth monotonic interpolating functions associated with monotonic data.

The literature offered some results and procedures [2,4] , and in more recent times (i.e. after this research had been completed) we also found [1,5] . However,none of these results fulfilled all of our requirements.

In this paper we present some new algorithms for performing this task, which are modifications and improvements of those stated in [2] . These algorithms show how to construct a piecewise linear, monotone interpolant, and then they use piecewise Bernstein polynomials in order to produce high order splines.

The algorithms are simple, and of the marching or local type. Thus, no system of equations have to be solved, and local knot adjustements can be made adaptively, for instance in order to inforce some bound restrictions on high order derivatives.

Even more recently Roulier [6] has also produced algorithms for shape preserving interpolation which are similar to the ones we offer.

## 2. PRELIMINARY RESULTS

In this section we shall give some definitions and results due to Mc Allister, Passow, and Roulier [2] , that will be used in the sequel. Let $\pi = \{x_0 < x_1 < \dots < x_N\}$ and let us consider the set of data pairs $\Delta = \{(x_i , y_i) , i=0,1,\dots,N\}$.

<u>Definition 2.1</u> The data in $\Delta$ are non-decreasing iff $y_i \leq y_{i+1}$ , $i=0,\dots,N-1$ . If all the inequalities are strict, then the data are increasing.

We consider now the successive slopes $M_i = (y_{i+1}-y_i)/(x_{i+1}-x_i)$ $i=0,\dots,N-1$ .

Definition 2.2 The data in $\Delta$ are non-concave iff $M_{i-1} \leq M_i$ , $i=1,\dots,N-1$. If all the inequalities are strict, then the data are convex. If the inequalities are reversed, then we say that the data are non-convex and concave, respectively.

We shall use the generic name of "monotone" for data that has either one of the properties : increasing, decreasing, non-increasing, or non-decreasing.

Let $0 < \alpha < 1$ , and consider the points $\bar{x}_i = x_{i-1} + \alpha h_i$ , with $h_i = x_i - x_{i-1}$ . Let $\bar{\pi} = \{\bar{x}_i\}$. We are interested in generating piecewise linear interpolants of the data $\Delta$, with breakpoints at the $\bar{x}_i$.

Definition 2.3 We shall say that the set of real numbers $\{\bar{y}_i\}$ , $i=1,\dots,N$ , is monotone $\alpha$-admissible, iff the piecewise linear function $L(x)$ constructed by joining the points $(x_0,y_0),(\bar{x}_1,\bar{y}_1),\dots,(\bar{x}_N,\bar{y}_N),(x_N,y_N)$ interpolates the data $\Delta$ and is itself monotone.

We will denote by $\sum_n^m(\pi)$ the set of splines of degree n and deficiency $n-m$ on $\pi$. That is $\varphi \in \sum_n^m(\pi)$ is a polynomial of degree at most $n$ in each subinterval $[x_{i-1}, x_i]$ and $\varphi \in C^m[x_0,x_N]$ . Passow and Roulier [4] have given conditions for the existence of $\alpha$-admissible sets. We state here a somewhat less general result than that of [4] , which will be sufficient for our purposes.

Theorem 2.4 Let $p,n$ be positive integers satisfying $p < n$ and let $\alpha = p/n$ . Let $m = \min(p, n-p)$. Then there exist sets $\{\bar{y}_i\}$ monotone (convex) $\alpha$-admissible for monotone (convex) data $\Delta = \{x_i,y_i\}$ iff for each $1 \leq k$ , there is a monotone (convex) spline $s \in \sum_{kn}^{km}(\pi)$ that interpolates the data $\Delta$ and that satisfies:
i) $s^{(j)}(x_i^+) = 0$ , $j=2,\dots,pk$ ; ii) $s^{(j)}(x_i^-) = 0$ , $j=2,\dots,(n-p)k$.

We observe then that for fixed even degree $n$, the maximum order of continuity is obtained when $p = n/2$ , which implies that $\alpha = 1/2$ . Since we are interested in smoothness for our interpolants, then we shall concentrate in this case from now on. All our results carry over to the more general case and even to the one considered in [2] .

Passow and Roulier also mention that, to obtain $s(x)$ constructively it is enough to consider $s(x) = S_i(x)$ on $[x_{i-1},x_i]$ , where $S_i(x)$ is the Bernstein polynomial of degree $kn$ associated with the restriction of the piecewise linear function $L(x)$ to

$[x_{i-1},x_i]$ . This is

(2.5) $$S_i(x)= \Big(\sum_{\nu=0}^{kn} L(x_{i-1}+\nu(x_i-x_{i-1})/kn)\binom{kn}{\nu}(x-x_{i-1})^{\nu}(x_i-x)^{kn-\nu}\Big)/(x_i-x_{i-1})^{kn}$$

3. MONOTONE INTERPOLATION

In addition to Theorem 2.4, Mc Allister et al [2] give an independent criterium (and constructive test) for the existence of $\alpha$-admissible sets associated with convex, increasing data. Unfortunately this $\alpha$ need not be 1/2, even for very innocent looking data. A further result says that $\alpha = 1/2$ can be obtained in the case that the third order differences of the data are non-negative.

In what follows we will show that if one drops the requirement of convexity in the resulting interpolant then he can obtain 1/2-admissibility, both for non-concave monotone and simply monotone data.

Theorem 3.1 Given a non-decreasing, non-concave set of data $\{x_i,y_i\}$, $i=0,\dots,N$ , there exists a non-decreasing piecewise linear function $L(x)$ that has break points at $\bar{x}_i = x_i + 0.5\,(x_{i+1}-x_1)$ and satisfies $L(x_i) = y_i$ .

Proof : Let $L_o(x) = M_o\,(x - x_1) + y_1$ and consider $L(x) = L_o(x)$ for $x_o \leq x \leq \bar{x}_1$ . Let $\bar{y}_1 = L_o(\bar{x}_1)$ . We define $L_i(x)$, the ith linear segment of $L(x)$, as the one joining the pair of points $(\bar{x}_i,\bar{y}_i)$, $(x_i,y_i)$ , i.e.

$$L_i(x) = (y_i - \bar{y}_i)(x - x_i)/(x_i - \bar{x}_i) + y_i$$

and take $L(x) = L_i(x)$ for $\bar{x}_i \leq x \leq \bar{x}_{i+1}$ . Finally, let $\bar{y}_{i+1} = L_i(\bar{x}_{i+1})$

Assume now that we have constructed successfully our polygonal through the (i-1)-segment. In order to continue the construction of the monotone interpolant we must have $y_i \leq \bar{y}_{i+1} \leq y_{i+1}$ .
It is easy to see that the worst case occurs for $\bar{y}_i = y_{i-1}$ . But then we would have

$$\bar{y}_{i+1} = L_i(\bar{x}_{i+1}) = (\bar{y}_i - y_i)(\bar{x}_{i+1} - x_i)/(\bar{x}_i - x_i) + y_i = M_{i-1}(x_{i+1} - x_i) + y_i .$$

But, because of the non-concavity of the data

$$\bar{y}_{i+1} = M_{i-1}(x_{i+1}-x_i) + y_i \leq M_i(x_{i+1}-x_i) + y_i = y_{i+1} .$$

Finally, since $\bar{y}_i \leq y_i$ , the slope of the ith segment is positive, and therefore it is obvious that $y_i \leq \bar{y}_{i+1}$ . □

The other combinations of monotonic behavior with convexity or concavity can all be reduced to the case just considered by simple changes of the dependent and/or independent variables. For instance, non-decreasing, non-convex data $(x_i,y_i)$ is transformed into the appropriate shape by considering $(-x_{n-i}, -y_{n-i})$.

If the data are non-concave (non-convex), it could happen that the construction of Theorem 3.1 fails. We show now how to modify the algorithm in the case that the data are only monotonic. The idea is to add some artificial data points.

Theorem 3.2 For any set of non-decreasing data points $(x_i,y_i)$, i=0,...,N , it is possible to construct a non-decreasing, piecewise linear interpolant with break points at $\bar{x}_i = x_i + 0.5\,(x_{i+1}-x_i)$.

Proof: The algorithm proceeds as in Theorem 3.1 , provided that $\bar{y}_{i+1} \leq y_{i+1}$. If this condition is violated, then we introduce one additional , auxiliary data point. Let us assume then , that $y_{i+1} < \bar{y}_{i+1}$. We introduce a new data pair $(x^*,y^*)$ with $x_i < x^* < \bar{x}_{i+1}$, $y_i < y^* < y_{i+1}$ , which , if chosen appropriately, will allow us to continue the construction of the monotone linear interpolant.

The auxiliary data point $x^*$ is defined as the abscissa of the intersection of the segment $L_i(x)$ with the line $y = y_{i+1}$ ,i.e. $x^* = (y_{i+1} - y_i)/ m_i + x_i$ , where the slope $m_i = (y_i-\bar{y}_i)/(x_i-\bar{x}_i)$. The new mid-points are

$$\bar{x}^* = 0.5\,(x^* + x_i) \quad , \quad \bar{x}^{**} = 0.5\,(x_{i+1} + x^*)$$

and the corresponding $\bar{y}^* = L_i(\bar{x}^*) = m_i\,(\bar{x}^* - x_i) + y_i < y_{i+1}$. We show now how to choose $y^*$ so that $\bar{y}^* \leq y^* \leq y_{i+1}$, and also that $\bar{y}^{**} \leq y_{i+1}$. In fact, any $y^*$ satisfying

$$\bar{y}^* \leq y^* \leq (x^* - \bar{x}^*)\,(y_{i+1}-\bar{y}^*)/(\bar{x}^{**} - \bar{x}^*)$$

will do , as is easily verified, and then the construction can be continued. □

We have implemented this procedures and they work well in our applications, as we shall report elsewhere [3] . It is also possible to produce shape preserving interpolants by using similar techniques.

## REFERENCES

1. Fritsch, F.N. and R.E. Carlson "Monotone piecewise cubic interpolation" . SIAM J. Numer. Anal. 17 : 238-246 (1980).
2. Mc Allister, D.F., E. Passow, and J.A. Roulier "Algorithms for computing slope preserving spline interpolation to data". Math.Comp. 31 : 717-725 (1977).
3. Pagallo, G. and V. Pereyra "Mesh selection by adaptive changes of variables". In preparation.
4. Passow, E.and J.A. Roulier "Monotone and convex spline interpolation". SIAM J. Numer. Anal. 14 : 904-909 (1977).
5. Pruess, S. "Alternatives to the exponential spline in tension".

Math. Comp. 33 : 1273-1281 (1979).

6. Roulier, J.A. "Constrained interpolation". SIAM J. Sci. Stat. Comp. 1 : 333-344(1980).

# SOME HYBRID IMPLICIT STIFFLY STABLE METHODS FOR ORDINARY DIFFERENTIAL EQUATIONS

ROLAND ENGLAND
IIMAS - UNAM
México 20, D.F.

## 1. INTRODUCTION

Most numerical methods for stiff initial value problems generate an algebraic approximation to the solution of the test equation $\dot{y}=\lambda y$, over a step h, which, for stiff stability, necessarily has at least one pole with $\mathrm{Re}(h\lambda) > 0$. Severe instability for problems with growing solutions is associated with a pole on the positive real axis - the most common situation. However, some methods, including many implicit Runge-Kutta methods have an even number of poles, all lying off the real axis, and affecting only growing oscillatory solutions, which are possibly less frequent in practical problems. In particular, the second derivative multistep methods of Enright have two complex conjugate poles, and other excellent stability properties: L-stability up to order 4, and stiff stability with strong stability at infinity up to order 9; and can satisfactorily represent purely oscillatory solutions for $|h\lambda| < 2$. Their disadvantage lies in the need to evaluate second derivatives. A class of hybrid implicit methods is introduced, with linear stability properties identical to those of the second derivative methods, but without the need to evaluate exact second derivatives.

## 2. LINEAR STABILITY ANALYSIS

Consider numerical step by step methods for the solution of a system of ordinary differential equations, written, for convenience, in autonomous form:

$$\dot{y} = F(y)$$

For standard linear stability analysis, it is supposed that the Jacobian matrix $\frac{\partial F}{\partial y}$ is equal to a constant matrix A, and frequently the remaining non-homogeneous forcing term is ignored, thus leaving the simple linear system of equations:

$$\dot{y} = Ay$$

Except in very exceptional cases, which are also ignored, the matrix A has a complete system of linearly independent eigenvectors $c_1, c_2, \ldots, c_N$, which form a matrix

$$C = [c_1, c_2, \ldots, c_N]$$

such that

$$AC = C\Lambda$$

where $\Lambda = \mathrm{diag}\{\lambda_1, \lambda_2, \ldots, \lambda_N\}$ is the diagonal matrix of the eigenvalues of A corresponding to the N eigenvectors $c_1, c_2, \ldots, c_N$. It is then possible to perform a change of variables, and introduce the new vector of dependent variables Z such that

$$Y = CZ$$

This vector Z satisfies the system of differential equations

$$\dot{Z} = C^{-1}ACZ = \Lambda Z$$

or written in terms of its elements

$$\dot{z}_i = \lambda_i z_i \quad , \quad i = 1, 2, \ldots, N.$$

Thus each element of Z satisfies a differential equation of the form

$$\dot{y} = \lambda y$$

independent of the other elements of the vector of dependent variables. However, the $\lambda$'s being eigenvalues of a general matrix with real valued elements, may themselves be complex constants.

Given a particular method, the selection of a suitable step size h for the original problem depends to a large extent on the eigenvalues $\lambda i$. The minimum absolute value $\min_{1 \leq i \leq N} |\lambda_i|$ is frequently the reciprocal of the longest time scale of interest for the problem, and so it is desirable that h be of the order of

$$1/ \min_{1 \leq i \leq N} |\lambda_i|$$

provided this is consistent with accuracy requirements.

For well posed initial value problems (stable with respect to the initial conditions) all the eigenvalues have negative real parts, and therefore

$$\left| \max_{1 \leq i \leq N} (\mathrm{Re}\lambda_i) \right| \leq \min_{1 \leq i \leq N} |\lambda_i|$$

For stiff initial value problems, which may be considered superstable, there are eigenvalues with large negative real parts, and

$$\max_{1\leqslant i\leqslant N}(-\mathrm{Re}\lambda_i) \gg \min_{1\leqslant i\leqslant N}|\lambda_i|$$

Consider now similar relations for well posed boundary value problems; there is no restriction on the sign of the real part of the eigenvalues, and so $\max_{1\leqslant i\leqslant N}(\mathrm{Re}\lambda_i)$ may also be larger than $\min_{1\leqslant i\leqslant N}|\lambda_i|$. For singular perturbation problems, which are an analogue of stiff initial value problems, there are typically large eigenvalues with positive and negative real parts of the same order of magnitud, and thus

$$\max_{1\leqslant i\leqslant N}(\mathrm{Re}\lambda_i) \sim \max_{1\leqslant i\leqslant N}(-\mathrm{Re}\lambda_i) \gg \min_{1\leqslant i\leqslant N}|\lambda_i|$$

Nevertheless, in practical situations, it would appear that the large eigenvalues are found most frequently close to the real axis, or close to the imaginary axis. Thus, rapidly oscillating solutions do not generally have strongly unstable growth either for well posed initial value problems or for most boundary value problems.

The analytic solution of the test equation $\dot{y}=\lambda y$ is $y(t)=y_0 e^{\lambda t}$. For some initial value $y_0$. Over one step of size h, defining $z=h\lambda$, the growth of the solution is given by

$$y(t+h) = y(t)\, e^{z}$$

A numerical method which is independent of the step size h, approximates this relation by

$$y_{s+1} = y_s R(z)$$

where, for multistep methods, $R(z)$ is a multiple valued function obtained as the root of an algebraic equation; the numerical solution is decomposed into a number of modes, each of which is multiplied by a different value of $R(z)$. The absolute stability region of the method is defined as:

$$\{z \in C|\text{ all values of } |R(z)| < 1\}$$

and the relative stability region as:

$$\{z \in C|\text{ all values of } |R(z)| < |e^{z}|\}$$

## 3. DESIRABLE PROPERTIES FOR A NUMERICAL METHOD

Many properties of a numerical method are reflected in its stability function R(z), and desirable properties of the method may be written as follows.

(1) Consistency of order $p \geqslant 1$ .

For some methods this is a complex algebraic condition (e.g. Runge-Kutta methods), but when applied to the test equation, reduces to:

one value of $R(z)=e^z+0(h^{p+1})$.

(2) Convergence also requires the property of asymptotic stability which is equivalent to:

all values of $|R(0)| \leqslant 1$, and only isolated values with $|R(0)| = 1$.

(3) For stiff initial value problems, stiff stability is defined by the conditions:

all values of $|R(z)| < 1$ for $\mathrm{Re}z < -D$ and for $|\mathrm{Im}z| < \theta$, $\mathrm{Re}z < 0$ , where for better stability, D is small and $\theta$ is large. Note that stiff stability implies $A(\alpha)$-stability with $\tan(\alpha) \geqslant \theta/D$.

(4) For very stiff initial value problems, damping of the solution for large negative $\lambda$ requires strong stability at infinity, defined by;

all values of $R(z) \to 0$ as $z \to -\infty$.

Note that as R(z) is algebraic, $R(z) \to 0$ as $|z| \to \infty$, there being only one limit value at infinity.

(5) For boundary value problems, relative stability near the positive real axis implies

all values of $|R(z)| < e^z$ for $z > 0$, $\mathrm{Im}z=0$ .

It is interesting to note that for the study of relative stability as defined in this paper, Nørsett's order stars (e.g. Wanner et al. 1978) are regions of relative instability.

## 4. AVAILABLE METHODS

Consider now to what extent existing methods fulfill the above desirable properties.

For one step methods, the stability function is a rational approximation to the exponential

$$R(z) = Q_1(z)/Q_0(z)$$

where $Q_0(z)$, $Q_1(z)$ are polynomials in z. For multistep methods, R(z) is given by the roots of the characteristic polynomial

$$Q(z,R) = Q_0(z)R^k + Q_1(z)R^{k-1} + \ldots + Q_k(z).$$

In this polynomial, for a k-step method, $Q_k(z)$ is not identically zero, and $Q_0(0) \neq 0$. The degree of $Q_k$ is $\alpha$, the number of zeros of $R(z)$, and the degree of $Q_0$ is $\omega$, the number of poles of $R(z)$.

Linear multistep methods have $\alpha \leqslant 1$, $\omega \leqslant 1$. On the other hand, stiff stability requires that the method be implicit, and therefore $\omega \geqslant 1$. In view of these conditions, it is natural to consider methods with $\omega=1$, of which the Backward Differentiation Formulae are a good example. Unfortunately, $\omega=1$ implies that $R(z)$ has one pole, which must lie on the positive real axis, and causes severe relative instability in its neighborhood.

Therefore, in order to fulfil properties (3) and (5) it is necessary to abandon linear multistep methods, and consider methods with $\omega > 1$, which implies methods with multiple stages, such as Implicit Runge Kutta methods, or which use higher derivatives in some way. Semi-implicit Runge-Kutta methods are also excluded, as they concentrate the poles of $R(z)$ on the positive real axis, often as one multiple pole. If all poles are to be kept away from the positive real axis, it is necessary that $\omega$ be even, as they occur in complex conjugate pairs.

The following are examples of methods with $\omega=2$.

1. Implicit Runge-Kutta methods of order 3:

(a)
$$y_{s+1/3} = y_s + \frac{5}{12} F(y_{s+1/3}) - \frac{1}{12} F(y_{s+1})$$
$$y_{s+1} = y_s + \frac{3}{4} F(y_{s+1/3}) + \frac{1}{4} F(y_{s+1})$$

(b)
$$y_{s+1/2} = y_s + \frac{1}{8} F(y_s) + \frac{1}{2} F(y_{s+1/2}) - \frac{1}{8} F(y_{s+1})$$
$$y_{s+1} = y_s + \frac{1}{6} F(y_s) + \frac{2}{3} F(y_{s+1/2}) + \frac{1}{6} F(y_{s+1})$$

Both are L-stable with stability function $R(z)$ equal to the Padé approximant

$$P_{12}(z) = (1+\frac{1}{3} z)/(1-\frac{2}{3} z + \frac{1}{6} z^2).$$

2. Implicit Runge-Kutta methods of order 4.

(a)
$$y_{s+(3-\sqrt{3})/6} = y_s + \frac{1}{4} F(y_{s+(3-\sqrt{3})/6}) + \frac{3-2\sqrt{3}}{12} F(y_{s+(3+\sqrt{3})/6})$$

$$y_{s+(3+\sqrt{3})/6} = y_s + \frac{3+2\sqrt{3}}{12} F(y_{s+(3-\sqrt{3})/6}) + \frac{1}{4} F(y_{s+(3+\sqrt{3})/6})$$

$$y_{s+1} = y_s + \frac{1}{2} F(y_{s+(3-\sqrt{3})/6}) + \frac{1}{2} F(y_{s+(3+\sqrt{3})/6})$$

(b) $$y_{s+1/2} = y_s + \frac{5}{24} F(y_s) + \frac{1}{3} F(y_{s+1/2}) - \frac{1}{24} F(y_{s+1})$$

$$y_{s+1} = y_s + \frac{1}{6} F(y_s) + \frac{2}{3} F(y_{s+1/2}) + \frac{1}{6} F(y_{s+1})$$

Both are A-stable with stability function $R(z)$ equal to the Padé approximant

$$P_{22}(z) = (1 + \frac{1}{2} z + \frac{1}{12} z^2)/(1 - \frac{1}{2} z + \frac{1}{12} z^2).$$

3. The Second Derivative Multistep methods of Enright (1974),

$$y_{s+1} = y_s + h \sum_{r=0}^{k} \beta_r \dot{y}_{s+1-r} + h^2 \gamma_0 \ddot{y}_{s+1}.$$

These methods are stiffly stable for oder $p \leqslant 9$ with the parameter $\theta$ taking the relatively large value of 2 for $p \leqslant 8$, and are L-stable for $p \leqslant 4$. They are all strongly stable at infinity, and relatively stable on the positive real axis, thus improving on the characteristics of the Implicit Runge Kutta methods given above. Their disadvantage lies in their need to calculate the second derivatives, and therefore the Jacobian matrix, at every step.

4. The Blended Linear Multistep methods of Skeel and Kong (1977).

Given the exact Jacobian matrix, these methods have linear stability properties identical to those of the Second Derivative methods. However, the Jacobian matrix is not required at every step, as an inexact approximation does not affect the order, but only modifies the stability regions. They require, on the other hand, twice as much storage, for past solution values in addition to derivative values.

## 5. ANOTHER APPROACH

Consider now the possibility of introducing a second stage, or off-step point, into multistep methods, rather than using second derivatives or the Jacobian matrix. Three of the above mentioned Implicit Runge Kutta methods may be obtained by using an interpolation formula using the unknown solution and its derivative at the forward point, evaluating the differential equation at some off-step point, and using the derivative in a quadrature rule to determine the unknown solution value (Hennart and England, 1979). Proceeding in the same way, consider the interpolation formula

$$y_{s+\theta} = \sum_{r=0}^{k} \alpha_r y_{s+1-r} + h\alpha \dot{y}_{s+1}$$

and the quadrature formula

$$y_{s+1} = y_s + h \sum_{r=0}^{k} b_r \dot{y}_{s+1-r} + h\beta F(y_{s+\theta})$$

On substituting $\dot{y} = F(y) = \lambda y$ and $z=h\lambda$, the characteristic polynomial is found to be

$$Q(z,R) = R^k - R^{k-1} - z \sum_{r=0}^{k} (b_r + \alpha_r \beta) R^{k-r} - z^2 \alpha\beta R^k$$

which is of the same form as that for the Second Derivative methods:

$$Q(z,R) = R^k - R^{k-1} - z \sum_{r=0}^{k} \beta_r R^{k-r} - z^2 \gamma_0 R^k$$

The conditions that the interpolation error be of order $h^P$ (global error), and the quadrature error be of order $h^{P+1}$ (Local error) are as follows:

$$\sum_{r=0}^{k} \alpha_r = 1 \quad , \quad \sum_{r=0}^{k} b_r + \beta = 1$$

$$\sum_{r=1}^{k} r\alpha_r - \alpha = 1-\theta \quad , \quad \sum_{r=1}^{k} rb_r + (1-\theta)\beta = \frac{1}{2}$$

$$\sum_{r=1}^{k} r^q \alpha_r = (1-\theta)^q \quad , \quad \sum_{r=1}^{k} r^q b_r + (1-\theta)^q \beta = \frac{1}{q+1} \quad , \quad q=2,3,\ldots,k+1=p-1$$

On eliminating the undetermined parameter $\theta$, these generate the following p conditions:

$$\sum_{r=0}^{k} b_r + \alpha_r \beta = 1$$

$$\sum_{r=1}^{k} r(b_r + \alpha_r \beta) - \alpha\beta = \frac{1}{2}$$

$$\sum_{r=1}^{k} r^q (b_r + \alpha_r \beta) = \frac{1}{q+1} \quad , \quad q=2,3,\ldots,k+1 = p-1$$

and comparing with the order conditions for the Second Derivative methods:

$$\sum_{r=0}^{k} \beta_r = 1$$

$$\sum_{r=1}^{k} r\beta_r - \gamma_0 = \frac{1}{2}$$

$$\sum_{r=1}^{k} r^q\beta_r = \frac{1}{q+1} , \quad q=2,3,\ldots,k+1 = p-1$$

it is seen that the coefficients must be related as follows:

$$\alpha\beta = \gamma_0$$

$$b_r + \alpha_r\beta = \beta_r , \quad r=0,1,\ldots,k$$

and that the characteristic polynomials of the corresponding methods are identical. The Hybrid Implicit Stiffly Stable methods (HImSti) thus have linear stability properties identical to those of the Second Derivative methods, and the Jacobian matrix is not needed, either to obtain the full order, or to give the stability regions.

However, the methods consist of two implicit equations to be solved simultaneously at every step. For non-stiff problems, a simple iterative substitution can obviously be used, but in the case of stiff problems, some other technique is needed in order to give convergence without restricting the step size. The Newton iteration matrix is quadratic in the Jacobian matrix, and in order to solve a system with such a matrix, it is necessary to adapt techniques used for two-stage implicit Runge-Kutta methods, or for the Second Derivative methods (Addison, 1979). Indeed, the method with k=1 is a third order implicit Runge-Kutta method.

The explicit solution of the order conditions, in terms of the undetermined parameter $\theta$, gives the coefficients of the methods as follows:

$$\alpha(\theta) = \frac{1}{k!} \prod_{r=0}^{k} (\theta+r-1)$$

$$\gamma_0 = \int_0^1 \alpha(\phi)\,d\phi$$

$$\beta(\theta) = \gamma_0/\alpha(\theta)$$

$$\left.\begin{aligned} \alpha_q(\theta) &= \frac{(-1)^{q-1}}{q \cdot q!\,(k-q)!}\,(\theta-1) \prod_{\substack{r=0 \\ r\neq q}}^{k} (\theta+r-1) \\ \beta_q &= \int_0^1 \alpha_q(\phi)\,d\phi \end{aligned}\right\} \quad q=1,2,\ldots,k$$

$$\alpha_0(\theta) = \frac{1}{k!}\left[1-(\theta-1)\sum_{q=1}^{k}\frac{1}{q}\right]\prod_{r=1}^{k}(\theta+r-1)$$

$$\beta_0 = \int_0^1 \alpha_0(\phi)\,d\phi$$

$$b_q(\theta) = \beta_q - \alpha_q(\theta)\beta(\theta)$$

$$= \frac{(-1)^{q-1}}{q!\,(k-q)!\,(\theta+q-1)} \int_0^1 (\phi-\theta) \prod_{\substack{r=0 \\ r\neq q}}^{k} (\phi+r-1)\,d\phi \quad , \quad q=0,1,\ldots,k$$

In particular, for k=1:

$$\alpha(\theta) = \theta(\theta-1) \quad , \quad \beta(\theta) = -1/6\theta(\theta-1)$$
$$\alpha_1(\theta) = (\theta-1)^2 \quad , \quad b_1(\theta) = (3\theta-1)/6\theta$$
$$\alpha_0(\theta) = -(\theta-2)\theta \quad , \quad b_0(\theta) = (3\theta-2)/6(\theta-1)$$

For k=2:

$$\alpha(\theta) = (\theta+1)\theta(\theta-1)/2 \quad , \quad \beta(\theta) = -1/4(\theta+1)\theta(\theta-1)$$
$$\alpha_2(\theta) = -(\theta-1)\theta(\theta-1)/4 \quad , \quad b_2(\theta) = -(2\theta-1)/24(\theta+1)$$
$$\alpha_1(\theta) = (\theta+1)(\theta-1)^2 \quad , \quad b_1(\theta) = (8\theta-3)/12\theta$$
$$\alpha_0(\theta) = -(3\theta-5)(\theta+1)\theta/4 \quad , \quad b_0(\theta) = (10\theta-7)/24(\theta-1)$$

For k=3:

$$\alpha(\theta) = (\theta+2)(\theta+1)\theta(\theta-1)/6$$
$$\alpha_3(\theta) = (\theta-1)(\theta+1)\theta(\theta-1)/18$$
$$\alpha_2(\theta) = -(\theta+2)(\theta-1)\theta(\theta-1)/4$$
$$\alpha_1(\theta) = (\theta+2)(\theta+1)(\theta-1)^2/2$$
$$\alpha_0(\theta) = -(11\theta-17)(\theta+2)(\theta+1)\theta/36$$
$$\beta(\theta) = -19/30(\theta+2)(\theta+1)\theta(\theta-1)$$
$$b_3(\theta) = (15\theta-8)/360(\theta+2)$$

$$b_2(\theta) = -(25\theta-13)/120(\theta+1)$$

$$b_1(\theta) = 19(5\theta-2)/120\theta$$

$$b_0(\theta) = (135\theta-97)/360(\theta-1)$$

These expressions have been calculated for all $k \leqslant 8$, the last case not being stiffly stable. They will be published at a later date, together with numerical results. To show their complexity, the coefficients for k=7 are as follows:

$$\alpha\ (\theta) = (\theta+6)(\theta+5)(\theta+4)(\theta+3)(\theta+2)(\theta+1)\theta(\theta-1)/5040$$

$$\alpha_7(\theta) = (\theta-1)(\theta+5)(\theta+4)(\theta+3)(\theta+2)(\theta+1)\theta(\theta-1)/35280$$

$$\alpha_6(\theta) = -(\theta+6)(\theta-1)(\theta+4)(\theta+3)(\theta+2)(\theta+1)\theta(\theta-1)/4320$$

$$\alpha_5(\theta) = (\theta+6)(\theta+5)(\theta-1)(\theta+3)(\theta+2)(\theta+1)\theta(\theta-1)/1200$$

$$\alpha_4(\theta) = -(\theta+6)(\theta+5)(\theta+4)(\theta-1)(\theta+2)(\theta+1)\theta(\theta-1)/576$$

$$\alpha_3(\theta) = (\theta+6)(\theta+5)(\theta+4)(\theta+3)(\theta-1)(\theta+1)\theta(\theta-1)/432$$

$$\alpha_2(\theta) = -(\theta+6)(\theta+5)(\theta+4)(\theta+3)(\theta+2)(\theta-1)\theta(\theta-1)/480$$

$$\alpha_1(\theta) = (\theta+6)(\theta+5)(\theta+4)(\theta+3)(\theta+2)(\theta+1)(\theta-1)^2/720$$

$$\alpha_0(\theta) = -(363\theta-503)(\theta+6)(\theta+5)(\theta+4)(\theta+3)(\theta+2)(\theta+1)\theta/705600$$

$$\beta\ (\theta) = -\ 33953/90(\theta+6)(\theta+5)(\theta+4)(\theta+3)(\theta+2)(\theta+1)\theta(\theta-1)$$

$$b_7(\theta) = (20625\theta-12062)/1814400(\theta+6)$$

$$b_6(\theta) = -(170265\theta-99359)/1814400(\theta+5)$$

$$b_5(\theta) = (207495\theta-120704)/604800(\theta+4)$$

$$b_4(\theta) = -(265641\theta-153761)/362880(\theta+3)$$

$$b_3(\theta) = (369399\theta-211886)/362880(\theta+2)$$

$$b_2(\theta) = -(608985\theta-341699)/604800(\theta+1)$$

$$b_1(\theta) = (2097735\theta-950684)/1814400\theta$$

$$b_0(\theta) = (551985\theta-416173)/1814400(\theta-1)$$

It remains to decide on a value of $\theta$ for each method in the family. For this purpose, it is of interest to note that they already require slightly less storage for past values than the Blended Linear Multistep methods, and that it is possible to reduce the storage requirement by one more derivative if it is chosen to set $b_k=0$. This choice implies

$$\theta = \frac{\int_0^1 \phi \prod_{r=0}^{k-1} (\phi+r-1)\,d\phi}{\int_0^1 \prod_{r=0}^{k-1} (\phi+r-1)\,d\phi}$$

a value which necessarily lies between 0 and 1, and is thus convenient for an off-step point. With this choice, the method for k=1 has coefficients:

$$\theta = 1/3 \;,\quad \alpha = -2/9 \;,\quad \beta = 3/4$$
$$\alpha_1 = 4/9 \;,\quad b_1 = 0$$
$$\alpha_0 = 5/9 \;,\quad b_0 = 1/4$$

and is exactly the Implicit Runge-Kutta method previously given as example (1a).

For k=2:

$$\theta = 1/2 \;,\quad \alpha = -3/16 \;,\quad \beta = 2/3$$
$$\alpha_2 = -1/32 \;,\quad b_2 = 0$$
$$\alpha_1 = 3/8 \;,\quad b_1 = 1/6$$
$$\alpha_0 = 21/32 \;,\quad b_0 = 1/6$$

For k=3:

$$\theta = 8/15 \;,\quad \alpha = -24472/151875 \;,\quad \beta = 3375/5152$$
$$\alpha_3 = 4508/455625 \;,\quad b_3 = 0$$
$$\alpha_2 = -3724/50625 \;,\quad b_2 = -1/552$$
$$\alpha_1 = 21413/50625 \;,\quad b_1 = 19/96$$
$$\alpha_0 = 291916/455625 \;,\quad b_0 = 25/168$$

For k=4,5,6,7, only the values of $\theta$, $\beta$ are presented here, as a sample of the rational coefficients which have all been calculated and will be published elsewhere:

k=4, $\theta = 21/38$ , $\beta = 19808792/30646665$

k=5, $\theta = 107/189$, $\beta = 2170458719541/3395869306240$

k=6, $\theta = 995/1726$,

$\beta = 57042106174536148730 72/90120546042257451820 05$

k=7, $\theta = 12062/20625$,

$$\beta = \frac{72767783072311431169509887695 3125}{11596241066993090954571465876 62336}$$

## REFERENCES

Addison, C.A. (1979). Implementing a Stiff Method Based upon the Second Derivative Formulas, Technical Report No. 130/79, University of Toronto, Department of Computer Science.

Enright, W.H. (1974). Second Derivative Multistep Methods for Stiff Ordinary Differential Equations, Vol. II, No. 2, SIAM Numer. Anal.

Hennart, J.P. and England, R. (1979). A Comparison Between Several Piecewise Continuous One Step Integration Techniques, pp. 33.1-33.4 in Working Papers for the 1979 SIGNUM Meeting on Numerical ODEs (Ed. R.D. Skeel), Report No. 963, University of Illinois at Urbana-Champaign, Department of Computer Science.

Skeel, R.D. and Kong, A.K. (1977). Blended Linear Multistep Methods, Vol. 3, ACM Trans. Math. Software.

Wanner, G., Hairer, E., and Nørsett, S.P. (1978). Order Stars and Stability Theorems, Vol. 18, BIT.

# Developing Effective Multistep Methods for the Numerical Solution of Systems of Second Order Initial Value Problems

by
W.H. Enright
Department of Computer Science
University of Toronto
M5S 1A7

ABSTRACT

In this paper various issues related to the development of effective software for second order initial value problems are discussed. In particular the difficulty caused by high frequency negligible components is identified and a class of multistep formulas suitable for problems with these components is proposed. The difficulties involved in choosing an appropriate error control strategy are reviewed and a particular strategy is proposed and justified.

## 1. Introduction

We are interested in developing numerical methods for second order initial value problems of the form

$$y'' = f(t,y,y'),\ y(t_o) = y_o,\ y'(t_o) = y'_o\ , \tag{1}$$

on the interval $t_o \le t \le t_F$. The ultimate goal of such an investigation is the production of variable-step, variable-order software packages that are comparable in their reliability and efficiency to those packages that have been developed for first order systems of initial value problems (see for example Shampine and Gordon (1975), Hindmarsh (1981) or Hull et. al. (1976)).

It is our belief that the recent substantial advancements in the development and distribution of reliable software for first order problems will be duplicated for second order problems.

In the case of first order problems the concept of 'stiffness' has been identified and analysed in great detail. The severe stepsize restriction imposed by the numerical stability of conventional methods when solving a stiff problem has led to the development of a special class of methods for these problems. An analogous situation arises in the numerical solution of second order equations where a special stability difficulty, closely related to stiffness, can arise and which necessitates the development of a special class of methods.

This special difficulty arises when the solution to the differential equation contains high frequency, low amplitude components. If the amplitude of these components is negligible, relative to the accuracy requirement, then these components do not have to be accurately reproduced in the numerical solution - they need only remain stable. The numerical stability of these negligible high frequency components will impose a severe stepsize restriction on conventional methods for second order problems. This difficulty is encountered for example in structural dynamics applications and often arises in the implicit form

$$My'' + Cy' + Ky = R(t) \quad (2)$$

The lack of suitable methods for this class of problems has led most researchers encountering such difficulties to employ fixed stepsize, low order formulas. To effectively use such techniques the user must have detailed knowledge of the formula he is using and the problem he is solving. In a recent investigation (Addison and Enright (1981)) this particular stability difficulty is discussed in detail and issues related to the design of variable-stepsize multistep methods for this class of problems are considered. It should be noted that, unlike the situation with stiff systems where the transient components are highly damped, these negligible high frequency components are generally at most lightly damped. Because of this a small change in the accuracy requirement or in the initial conditions of a second order problem can result in a high frequency component becoming significant and hence the cost of the integration can be extremely sensitive to these parameters. (When these components are significant they must be accurately reproduced and this implies a restriction on the stepsize.) Note that this sensitivity is inherent in the mathematics of the problem specification and does not arise from numerical considerations. On the other hand, the error estimator of a reliable numerical method should be able to implicitly identify when a high frequency component is significant and restrict the stepsize accordingly.

There are two approaches one can follow to obtain a numerical solution of a second order problem. One can rewrite the system as a larger system of first order equation and then use a (possibly modified) first order package to obtain the numerical solution. This approach has proved to be effective for most second order problems. If the difficulty of negligible high frequency components is present then one must use a first order method designed for stiff systems. The second approach is to use methods designed specifically for second order problems.

Example of such direct methods are the multistep method developed by Krogh (1971) and Runge Kutta formula pairs developed and analysed by Bettis and Horn (1978). Investigations (Krogh (1975), Bettis and Horn (1978)) have shown that for the general problem (1) the performance of such direct methods is comparable in efficiency to that of the first approach and in the special case where y' does not appear in the differential equation the performance of such direct methods is superior. We have been particularly interested in investigating whether direct multistep methods are suitable for problems with negligible high frequency components. This question will be considered in the next section where the concept of a formula-pair is introduced.

A critical constituent of any initial value method is the local error control strategy. This is the strategy which determines whether a given approximation is acceptable and what the next attempted stepsize will be. In the case of first order systems it is now generally accepted that one must keep the magnitude of the local-error-per-unit-step less than a multiple of the tolerance (although this is frequently accomplished by keeping the local-error-per-step less than the tolerance and performing local extrapolation on each step). When one is considering direct methods for second order problems one must also be very careful about the choice of error control strategy. In section three we will address this question and justify a particular strategy. We will also attempt to indicate areas for future work.

## 2. Formula-pairs for second order problems.

In Addison and Enright (1981) the concept of a formula-pair is introduced as a basis for direct multistep methods for second order problems. A formula-pair is defined by

$$(3)\qquad \begin{aligned} y_{i+1} &= \sum_{j=1}^{k} \alpha_j y_{i+1-j} + h \sum_{j=0}^{k} \beta_j y'_{i+1-j} \\ &\qquad + h^2 \sum_{j=0}^{k} \gamma_j y''_{i+1-j} \\ y'_{i+1} &= \sum_{j=1}^{k} \bar{\alpha}_j y'_{i+1-j} + h \sum_{j=0}^{k} \bar{\beta}_j y''_{i+1-j} \,. \end{aligned}$$

Note that the $\alpha$'s, $\beta$'s and $\gamma$'s can be viewed as a second derivative multistep formula for first order problems and the $\bar{\alpha}$'s and $\bar{\beta}$'s can be viewed as a standard multistep formula for first order problems. If one develops direct methods for second order problems based on this class of formula-pairs then a nonlinear system of equations must be solved on each time step (for the unknown $y_{i+1}$ and $y'_{i+1}$). The 5k+3 parameters required to specify a formula-pair can be chosen to ensure desirable stability and accuracy properties of the resulting method. It should also be noted that in the special case that $\alpha_j=\overline{\alpha}_j$ for j=1, 2 ... k; $\beta_j=\overline{\beta}_j$ for j=0,1 ... k; and $\gamma_j=0$ for j=0,1 ... k this formula-pair is equivalent to the approach of rewriting the problem as an equivalent first order problem and applying the corresponding multistep formula (determined by the $\alpha$'s and $\beta$'s). In Addison and Enright (1981) the stability and order of formula-pairs are investigated in detail and formulas suitable for problems with high frequency negligible components are identified. It is shown that the formula pair will be of order p if the $\alpha$'s, $\beta$'s and $\gamma$'s correspond to a second derivative multistep formula of order p and the $\overline{\alpha}$'s and $\overline{\beta}$'s correspond to a multistep formula of order p.

The stability of the formula pair is more difficult to analyse and requires some preliminary definitions and the introduction of an appropriate model problem. If the formula-pair is applied to the second order problem

$$y'' + Cy' + Ky = 0 \ , \qquad y(t_o) = y_o,\ y'(t_o) = y'_o \ , \tag{4}$$

where C and K are n×n matrices, then the formula-pair will be stable for a given fixed stepsize k if the resulting approximation $y_i$ and $y'_{i+1}$ are well defined and uniformly bounded in norm for all i.

If we follow standard notation and let $\alpha(\omega) = \omega^k - \sum_{j=1}^{k} \alpha_j \omega^{k-j}$, $\beta(\omega) = \sum_{j=0}^{k} \beta_j \omega^{k-j}$ and $\gamma(\omega) = \sum_{j=0}^{k} \gamma_j \omega^{k-j}$ with $\overline{\alpha}(\omega)$ and $\overline{\beta}(\omega)$ defined similarly then we can introduce the polynomials

$$q_1(\omega) = -\gamma(\omega)\overline{\alpha}(\omega) + \beta(\omega)\overline{\beta}(\omega) \ ,$$
$$q_2(\omega) = \alpha(\omega)\overline{\beta}(\omega) \ ,$$
$$\text{and} \qquad q_3(\omega) = \alpha(\omega)\overline{\alpha}(\omega) \ .$$

Addison and Enright (1981) show that a formula-pair applied to the model system (4) with stepsize h will be stable if and only if whenever $\omega$ is a root of

$$\det[h^2 q_1(\omega)K + hq_2(\omega)C + q_3(\omega)I] = 0$$

then the magnitude of $\omega$ will be less than or equal to one.

This characterization of the numerical stability of a formula pair is then used to associate a region of stability with a formula-pair and formula-pairs are investigated and compared on the basis of their regions of stability. In particular it is shown that a formula-pair can have a very restricted region of stability even though both constituent formulas, when viewed as formulas for first order problems, have large regions of stability.

This characterization of the order and stability of a formula-pair provides a convenient tool for determining whether a particular formula-pair is a suitable candidate for implementation as a direct method for second order problems. It also provides an analytical measure that can be used to quantify the potential advantages that a formula-pair may possess over the alternative approach of solving the equivalent first order problem with a standard multistep formula. A preliminary investigation (Addison (1980)) seems to indicate that, from the stability and accuracy viewpoint, the best formula-pair that has been developed is not significantly better than the best single multistep formula.

## 3. Error control

Several issues must be considered when one is developing error and stepsize control strategies for a method for second order initial value problems. As in the case for first order problems the goal of these strategies is to monitor and control the local error in such a way that the chosen stepsize is as large as possible subject to the condition that the magnitude of the global error will be comparable to the prescribed error tolerance. Most methods for second order problems generate approximations to y and y' on each time step and one must decide whether the error control should monitor both of these approximations. In addition one must decide what an appropriate bound for the local error should be to ensure that the magnitude of the global error is comparable to the error tolerance. Before we analyse these questions in more detail we will review the relevant analysis and strategies for methods for first order problems and this will suggest alternatives for second order methods.

In the analysis of the relationship between local and global error for the first order initial value problem,

$$y' = f(t,y), \; y(t_o) = y_o, \tag{5}$$

one has the classical result that, for a method based on a multistep formula with a fixed stepsize, if the local error is $0(h^{p+1})$ the global error will be $0(h^p)$. When a variable-step method is analysed it is often more convenient to introduce the 'defect' of the numerical method and relate the magnitude of the defect to the magnitude of the local error. Results from the theory of perturbed ODEs can then be used to relate the magnitude of the defect to the magnitude of the global error. One can show that, if TOL is the prescribed error tolerance and the local error on each time step is bounded by TOL/h, then the computed approximations $y_i$ (for i=0,1 ... N) will lie on the exact solution z(t) to the perturbed initial value problem

$$z' = f(t,z) + \delta(t), \quad z(t_o) = y_o,$$

where $\|\delta(t)\|$ is bounded by a small multiple of TOL. The vector function $\delta(t)$ is called the defect of the approximation z(t). Standard results can then be used to show that the magnitude of the global error, y(t)-z(t), will also be bounded by a multiple of TOL (this multiple depending only on the equation and not on the method or stepsizes used). One of the most popular local error control strategies for first order systems is based on attempting to indirectly bound the local-error-per-unit-step by a small multiple of TOL by estimating and directly controlling the local-error-per-step of the unextrapolated solution and then performing local extrapolation to obtain the final approximation.

A similar analysis of the relationship between local and global errors for second order problems can be applied and this will suggest a particular error control strategy for this class of problems. The corresponding single formula fixed stepsize result is that if the local error in y is $O(h^{p+2})$, then the global error is $O(h^p)$. At first glance this would suggest that a tighter local error control is required for second order problems than is the case for first order problems. On the other hand, we have seen from the last section that if approximations are generated by a formula-pair to both y and y' and if both of these approximations have local errors $O(h^{p+1})$ then the global error of the approximation will be $O(h^p)$. To analyse variable-stepsize methods one can introduce the concept of a defect for methods which approximate both y and y' when applied to (1). If the local error on each time step for both y and y' are bounded by TOL/h then one can show that there exists z(t), $\bar{z}(t)$ twice differentiable at all but a finite number of points in $t_o, t_N$ satisfying:

i) $z'' = f(t,z,\bar{z}) + \delta_1(t)$,

ii) $\bar{z}(t) = z'(t) + \delta_2(t)$,

iii) $z(t_i) = y_i$, $\bar{z}(t_i) = y_i'$ for i=0,1 ... N,

where $\|\delta_1(t)\|$ and $\|\delta_2(t)\|$ are both bounded by a small multiple of TOL. Differential inequalities can then be used to show that the global error in both y and y' will be bounded by a multiple of TOL. This suggests that a reliable error control strategy can be developed based on estimating and controlling the local-error-per-unit-step in the approximations to both y and y'. It should be noted that the conditions we have developed here have only been shown to be sufficient and it may be possible to relax the accuracy requirement for the approximation to y'.

(This would require an appropriate modification of the notion of a defect and an accompanying result relating the global error to the defect.)

It is clear that more work remains to be done in the development of software for second order problems. Methods need to be developed and compared for both standard problems and problems with high frequency negligible components. While we have concentrated on multistep methods in this paper, methods based on extrapolation and Runge-Kutta-type formulas merit similar investigations and a library of methods should be collected. Some direct methods do not approximate y' explicitly and a suitable error control for these methods must also be implemented and justified. We feel that such investigations, combined with extensive numerical testing, will lead not only to more efficient and reliable methods but also to a better understanding of the behaviour of numerical methods for initial value problems.

References


C.A. Addison (1980), Numerical methods for a class of second order ODEs arising in structural dynamics, Ph.D. Thesis, appeared as Dept. Comp. Sc. Tech. Rep. No. 147/80, University of Toronto, Toronto.

C.A. Addison and W.H. Enright (1981), Properties of multistep formulas intended for a class of second order ODEs, submitted to SIAM J. Numer. Anal.

D.G. Bettis and M.K. Horn (1978), Embedded Runge-Kutta-Nystrom algorithms of order two through six, TICOM Rep. No. 78-3, University of Texas, Austin.

A.C. Hindmarsh (1980), LSODE and LSODEI, two new initial value ordinary differential equation solvers, ACM SIGNUM Newsletter, 15, pp. 10-12.

T.E. Hull, W.H. Enright and K.R. Jackson (1976), Users guide for DVERK - a subroutine for solving non-stiff ODEs, Dept. of Comp. Sc. Tech. Rep. No. 100, University of Toronto, Toronto.

F.T. Krogh (1971), Suggestion on conversion (with listings) of the variable order integrators VODQ, SVDA and DVDQ, J.P.L. Tech. Memo. 278, Jet Propulsion Laboratory, CIT, Pasadena.

F.T. Krogh (1975), Summary of test results with variants of a variable order Adams method, Computing Memo. 376, Section 914, Jet Propulsion Laboratory, CIT, Pasadena.

L.F. Shampine and M.K. Gordon (1975), Computer Solution of Ordinary Differential Equations, W.H. Freeman, San Francisco.

# NUMERICAL SOLUTION OF SINGULAR TWO-POINT BOUNDARY-VALUE PROBLEMS BY INVARIANT IMBEDDING

P. Nelson and S. Sagong
Institute for Numerical Transport Theory
Department of Mathematics
Texas Tech University
Lubbock, Texas 79409/USA

and

I. T. Elder
Department of Mathematics, Computer
Science and Statistics
Eastern New Mexico University
Portales, New Mexico 88130/USA

## 1. Introduction

The method of invariant imbedding has been widely used [1-3] in many contexts, but little effort seems to have been devoted to its application to singular differential systems. Scott [1, Sec.V.8] considered sufficient conditions for existence of two different versions of invariant imbedding (the "r" and "s" equations) for linear second-order equations having a regular singular point. This work constitutes a significant extension of earlier results of Banks and Kurowski [4]. More recently, Elder [5] showed how to apply the subscheme of invariant imbedding known as integration-to-blowup [6,7] to compute the smallest eigenlength of a linear first-order system having a singularity of the first kind. The purpose of the present paper is to show how the approach of Elder can be adapted to the solution of certain singular two-point boundary-value problems by means of Scott's version [1,8-10] of invariant imbedding. Specifically, the approach applies to problems based on a linear homogeneous first-order differential system with a singularity of the first kind [11, Sec. 4.2] and boundary conditions consisting of existence (finite!) at the singularity and specified values at some second point.

## 2. Preliminaries

We wish to obtain solutions of a linear homogeneous first-order differential system with a singularity of the first kind at $z = 0$,

$$\underset{\sim}{u}'(z) = \left(\frac{A_0}{z} + A_1(z)\right)\underset{\sim}{u}(z) + \left(\frac{B_0}{z} + B_1(z)\right)\underset{\sim}{v}(z), \tag{1a}$$

$$\underset{\sim}{v}'(z) = \left(\frac{C_0}{z} + C_1(z)\right)\underset{\sim}{u}(z) + \left(\frac{D_0}{z} + D_1(z)\right)\underset{\sim}{v}(z), \tag{1b}$$

subject to boundary conditions of the form

$$\underset{\sim}{U}(0+) = \underset{\sim}{M}, \underset{\sim}{V}(0+) = \underset{\sim}{L} \; ; \; \underset{\sim}{V}(x) = \underset{\sim}{\beta}, \; x > 0. \tag{2a,b}$$

Here $\underset{\sim}{U}$ and $\underset{\sim}{V}$ are respectively an m- and an n- vector of dependent variables, M and L are vectors having finite component variables and $A_i$, $B_i$, $C_i$, $D_i$, for i = 0,1 are matrices of the appropriate dimensions (e.g. $A_i$ is $m \times m$). The matrices with subscript 0 are constant and those with subscript 1 are analytic on $(0,\delta]$ for some $\delta > 0$ and piecewise continuous on $[0,\infty)$.

Theorem A (Elder [5]): *Let an $(m + n) \times q$ matrix*

$$Y(z) = \begin{pmatrix} Y_1(z) \\ Y_2(z) \end{pmatrix} \tag{3}$$

*be a fundamental matrix for (1) - (2a), where $Y_1$ and $Y_2$ have respectively m and n rows. Define the $(m + n) \times (m + n)$ matrix $E_0$ by*

$$E_0 = \begin{pmatrix} A_0 & B_0 \\ C_0 & D_0 \end{pmatrix} . \tag{4}$$

*Let $H^+$ be the vector space spanned by the eigenvectors and generalized eigenvectors of $E_0$ associated with eigenvalues $\lambda$ satisfying $Re(\lambda) > 0$ along with the proper eigenvectors of $E_0$ associated with eigenvalue $\lambda = 0$. Let*

$$H = \begin{pmatrix} H_m^+ \\ H_n^+ \end{pmatrix} \tag{5}$$

*be any $(m + n) \times n$ matrix whose columns constitute a basis for $H^+$ with the subscripts representing the number of rows of each submatrix.*

*Then the following are true.*

*(i) $q = \dim H^+$.*

*(ii) The TPBVP (1)-(2) has a unique solution iff $q = n$ and $Y_2(x)$ is invertible.*

*(iii) If $q = n$, then $Y_2(x)$ is invertible for all sufficiently small positive x iff $H_n^+$ is invertible.*

*(iv) If (1)-(2) has a unique solution for every x satisfying $0 < x < X$ for some $X > 0$, then the matrix*

$$R(z) \equiv Y_1(z)Y_2(z)^{-1} \tag{6}$$

*satisfies the Riccati differential equation*

$$R'(z) = \left(\frac{B_0}{z} + B_1(z)\right) + \left(\frac{A_0}{z} + A_1(z)\right) R(z) - R(z)\left(\frac{D_0}{z} + D_1(z)\right)$$
$$- R(z)\left(\frac{C_0}{z} + C_1(z)\right) R(z) \tag{7}$$

*and*

$$R(0^+) = H_m^+ (H_n^+)^{-1}. \tag{8}$$

(v) $Y_2(0)$ *is invertible iff* $\mathcal{H}^+$ *consists entirely of eigenvectors of* $E_0$ *associated with the eigenvalue* $\lambda = 0$.

Remark: Henceforce we assume the existence of $\mathcal{X}$ as described in (iv).

Now we will define the basic invariant imbedding quantities R and T for first-order systems, formulate initial value procedures for determination of R and T by numerical solution of initial-value problems, and express the solution of the BVP in terms of these quantities.

## 3. Invariant Imbedding for First-Order Systems

Let $\underset{\sim}{u}, \underset{\sim}{v}$ be any solution of the TPBVP (1)-(2a). Then it is readily shown that there exists a constant vector $\underset{\sim}{c}$ s.t.

$$\underset{\sim}{u}(z) = Y_1(z)\underset{\sim}{c} \; ; \; \underset{\sim}{v}(z) = Y_2(z)\underset{\sim}{c} . \tag{9a,b}$$

If $\underset{\sim}{v}$ satisfies (2b) as well, then

$$\underset{\sim}{c} = Y_2(x)^{-1}\underset{\sim}{\beta} . \tag{10}$$

Define T by

$$T(z) = Y_2(z)^{-1}. \tag{11}$$

It then follows from (9) that

$$\underset{\sim}{v}(z) = T(z)^{-1}T(x)\underset{\sim}{\beta} \; ; \; \underset{\sim}{u}(z) = R(z)\underset{\sim}{v}(z). \tag{12a,b}$$

Furthermore, T satisfies the differential system

$$T'(z) = -T(z)\left[\left(\frac{D_0}{z} + D_1(z)\right) + \left(\frac{C_0}{z} + C_1(z)\right) R(z)\right]. \tag{13}$$

As (7) and (8) comprise an initial value problem for R, we only need to determine the value of T(0) in order to reduce solution of the TPBVP (1)-(2) to that of initial-value problems. Then T can be obtained by numerical solution of the resulting initial-value problem along with that for R.

If $Y_2(0)$ is nonsingular, we can simply take T(0) as any invertible $n \times n$ matrix. Otherwise we choose a small positive number $z_0$, and determine $Y_2(z)$ on $[0,z_0]$ by numerical integration of (7), (8) and

$$Y_2'(z) = \left[\left(\frac{C_0}{z} + C_1(z)\right) R(z) + \left(\frac{D_0}{z} + D_1(z)\right)\right] Y_2(z), \tag{14}$$

subject to

$$Y_2(0) = P_2, \tag{15}$$

where $P = (P_1^T \;\; P_2^T)^T$ is any $(m + n) \times n$ matrix the nonzero columns of which are a basis for the null space of $E_0$. We then set $T(z_0) = Y_2(z_0)^{-1}$, and obtain $R(z)$, $T(z)$ for $z_0 \leq z \leq x$ by numerical integration of (7) and (13) subject to the known values of $R(z_0)$ and $T(z_0)$. The required values of $\underset{\sim}{u}(z)$ and $\underset{\sim}{v}(z)$ can then be computed by (12).

There remains one further computational detail. Typical initial value codes require the user to supply a subroutine which computes the derivative. However, the right-hand side of (7) is indeterminate at $z = 0$ for $R(0)$ given by (8) and (5), and likewise the right-hand side of (13) in the case that $T(0)$ exists. For the problems we have encountered it has been possible to determine $R'(0)$ (and $T'(0)$, when $T(0)$ exists) by applying L'Hospital's rule to (7) (or (13)) although we have no general proof of the effectiveness of this approach.

We now apply the above results to second-order equations.

## 4. Single Second-Order Equations

We are interested in solving a general scalar homogeneous second-order differential equation having a singular point at $z = 0$,

$$y''(z) + \left(\frac{a_0}{z} + a_1(z)\right) y'(z) + \left(\frac{b_0}{z^2} + \frac{b_1(z)}{z}\right) y(z) = 0, \tag{16}$$

subject to

$$y(0+) = M(|M| \neq \infty) \;\; ; \;\; y(x) = \beta (x > 0). \tag{17}$$

Here $a_0$ and $b_0$ are real constants, while $a_1(z)$ and $b_1(z)$ are assumed to satisfy conditions similar to those imposed on $A_1(z)$ etc. in Section 2.

We convert (16) under the transformation of dependent variables

$$u(z) = zy'(z) \;\; ; \;\; v(z) = y(z) \tag{18}$$

to obtain an equivalent system of the form (1), namely

$$u'(z) = \left(\frac{1-a_0}{z} - a_1(z)\right) u(z) - \left(\frac{b_0}{z} + b_1(z)\right) v(z), \tag{19a}$$

$$v'(z) = \frac{u(z)}{z}. \tag{19b}$$

It also follows from Theorem 5.1 of [11] that $u(0+) = \lim_{z \to 0+} zy'(z)$ exists iff $v(0+) = y(0+)$ exists. Thus, the boundary condition (17a) is equivalent to (2a) under the transformations (18), and the results of the preceding section are directly applicable to (19) and the corresponding transformed boundary conditions.

By comparing (1) and (19) we identify $A_0 = 1 - a_0$, $A_1 = -a_1$, $B_0 = -b_0$, $B_1 = -b_1$, $C_0 = 1$, $C_1 = D_0 = D_1 = 0$, and the matrix $E_0$ is

$$E_0 = \begin{pmatrix} 1-a_0 & -b_0 \\ 1 & 0 \end{pmatrix} . \tag{4'}$$

The equations corresponding to (7), (12), (13), and (14) are respectively

$$r'(z) = \left(-\frac{b_0}{z} - b_1(z)\right) + \left(\frac{1-a_0}{z} - a_1(z)\right) r(z) - \frac{r^2(z)}{z} , \tag{7'}$$

$$y(z) = \frac{t(x)}{t(z)} \beta \ ; \quad zy'(z) = r(z)y(z), \tag{12'a,b}$$

$$t'(z) = -\frac{t(z)r(z)}{z} , \tag{13'}$$

$$y_2'(z) = \frac{y_2(z)r(z)}{z} . \tag{14}$$

Note that (7') is different from either the s-equation or the r-equation of Scott [1, Sec.V.8], although it is quite similar to the s-equation. (For us $r = zy'/y$, whereas Scott's s is given by $s = y'/y$.)

As all components of $E_0$ are real, its eigenvalues are either real of comprise a pair of complex conjugates. Thus we conclude that the TPBVP (16)-(17) has a unique solution for all sufficiently small x iff one of the following conditions holds:

(i) The matrix $E_0$ has one positive and one negative eigenvalue, and the eigenvector associated with the positive eigenvalue has nonzero second component.

(ii) The value $\lambda = 0$ is an eigenvalue of $E_0$ with geometric multiplicity one, and the other eigenvalue (if any) is negative.

These are readily shown to be respectively equivalent to

(i') $b_0 < 0$ and

(ii') $b_0 = 0$ and $a_0 \geq 1$.

In either case, the appropriate value of r(0) is the ratio of the first component to the second component of the eigenvector associated with the positive eigenvalue or with zero, as the case may be.

If t(0) exists (i.e. $y(0) \neq 0$), which is precisely the case of (ii') (or (ii)) above, then r and t can be determined by numerical integration of (7'), (13') subject to the initial value r(0) described above and any nonzero value for t(0). Then y can be determined by (12'a).

If t(0) does not exist (i.e. y(0) = 0), which is precisely the case of (i')(or (i)) above, then $y_2(z_0)$ and $r(z_0)$ should be determined by numerical integration of (7'), (14') subject to R(0) as described above and $y_2(0) = 0$. Then r and t are determined by numerical integration of (7'), (13') subject to the known values of $r(z_0)$ and $t(z_0) = 1/y_2(z_0)$. Then y is determined from (12'a).

The above procedures are illustrated in the following two examples: one for the case when $y(0) \neq 0$ and the other when $y(0) = 0$.

## 5. Numerical Examples

Example 1.

This example consists of the scalar second-order equation

$$y''(z) + \frac{y'(z)}{z} + y(z) = 0, \tag{16-1}$$

subject to

$$y(0+) = M(|M| \neq \infty); \quad y(1) = J_0(1) \tag{17-1a,b}$$

where $J_0$ is the Bessel function of first kind of order zero.

By comparing (17) and (17-1), we see that $a_0 = 1$, $a_1(z) = 0$, $b_0 = 0$, $b_1(z) = z$, and hence we are in case (ii'); i.e. $y(0) \neq 0$, so that t(0) exists.

The equation corresponding to (7') is

$$r'(z) = - z - \frac{r^2(z)}{z}, \tag{7-1}$$

and the matrix $E_0$ is

$$E_0 = \begin{pmatrix} 0 & 0 \\ 1 & 0 \end{pmatrix}, \tag{4-1}$$

which has eigenvalue $\lambda = 0$. The eigenvector of $E_0$ corresponding to $\lambda = 0$ is $(0\ 1)^T$. Thus $r(0) = 0$, and t(0) can be taken as any nonzero number. The right-hand sides of (7-1) and (13') are indeterminate at $z = 0$, but one application of L'Hospital's rule gives $r'(0) = t'(0) = 0$.

The solution of (16-1)-(17-1) is $y(z) = J_0(z)$. In Table I we compare known values of $J_0(z)$ with y(z) as obtained from (12'a) with $\beta = J_0(1)$, where t is determined by numerical integration of (7-1), (13') subject to $r(0) = 0$, $t(0) = 1$ and $r'(0) = t'(0) = 0$.

TABLE I - RESULTS FOR EXAMPLE 1

| z | y(z) | $J_0(z)$ | z | | |
|---|---|---|---|---|---|
| 0.0 | 0.99999995 | 1.00000000 | 0.6 | 0.91200490 | 0.91200486 |
| 0.1 | 0.997500159 | 0.997501562 | 0.7 | 0.88120091 | 0.88120089 |
| 0.2 | 0.99002501 | 0.99002497 | 0.8 | 0.84628737 | 0.84628735 |
| 0.3 | 0.97762629 | 0.97762625 | 0.9 | 0.80752381 | 0.80752380 |
| 0.4 | 0.96039827 | 0.96039823 | 1.0 | $= J_0(1)$ | 0.765197686557... |
| 0.5 | 0.93846984 | 0.93846981 | | | |

Example 2.

This example consists of the scalar second-order equation

$$y''(z) + \frac{y'(z)}{z} + \left(1 - \frac{1}{z^2}\right) y(z) = 0 \tag{16-2}$$

subject to

$$y(0+) = M_2 \quad (|M_2| \neq \infty); \; y(I) = J_1(1), \tag{17-a,b}$$

where $J_1$ is the Bessel function of first kind of order 1.

By comparing (16) and (16-2), we see that $a_0 = 1$, $a_1(z) = 0$, $b_0 = -1$, $b_1(z) = z$, and hence we are in case (i'); i.e. $y(0) = 0$, so that $t(0)$ is not defined.

The equation corresponding to (7') is

$$r'(z) = \frac{1}{z} - z - \frac{r^2(z)}{z} \tag{7-2}$$

and the matrix $E_0$ is

$$E_0 = \begin{pmatrix} 0 & 1 \\ 1 & 0 \end{pmatrix}, \tag{4-2}$$

The eigenvector associated with $\lambda = +1$ is $(1\ 1)^T$, so that $r(0) = 1$. The right-hand sides of (7-2) and (14') are indeterminate at $z = 0$, and one application of L'Hospital's rule gives $r'(0) = 0$. One can take $y_2'(0)$ as any nonzero number (we used $y_2'(0) = 1$), because $y_2'(0)$ is merely a scale factor which ultimately cancels in (12'a).

The solution of (16-2)-(17-2) is $y(z) = J_1(z)$. In Table II we compare known values of $J_1(z)$ with $y(z)$ as obtained from (12'a) with $\beta = J_1(1)$, where t is determined by the second method; that is, with $z_0 = 0.1$, $r(z_0)$ and $y_2(z_0)$ are determined by integration of (7-2), (14') subject to $r(0) = 1$, $y_2(0) = 0$, $r'(0) = 0$, $y_2'(0) = 1$ from $z = 0$ to $z = z_0 = 0.1$ and then t and r are determined by integration of (7-2), (13') subject to $r(z_0)$ and $t(z_0) = 1/y_2(z_0)$.

TABLE II - RESULTS FOR EXAMPLE 2

| $z$ | $y(z)$ | $J_1(z)$ | $z$ | | |
|---|---|---|---|---|---|
| 0.0 | 0.0* | 0.0 | 0.6 | 0.286700985 | 0.286700988 |
| 0.1 | 0.04993757 | 0.04993753 | 0.7 | 0.328995737 | 0.328995742 |
| 0.2 | 0.09950092 | 0.09950083 | 0.8 | 0.368842042 | 0.368842046 |
| 0.3 | 0.14831885 | 0.14831882 | 0.9 | 0.405949544 | 0.405949546 |
| 0.4 | 0.19602659 | 0.19602658 | 1.0 | $\equiv J_1(1)$ | 0.440050585744... |
| 0.5 | 0.242268459 | 0.242268458 | | | |

*To avoid divide check from $t(0) = 1/y_2(0) = 1/0$, $v(0) = t(1)y_2(0)J_1(1)$ is used instead of $v(0) = t(1)\cdot J_1(1)/t(0)$, which is the exact form of (12'a).

For both examples, integrations were effected by means of the RKF code of Shampine and Allen [12] in double precision on the ITEL AS/6 at Texas Tech University. Error tolerances (relative and absolute) of $10^{-4}$ were used.

## 6. Concluding Remarks

The preliminary results in this note show that it is feasible to solve singular homogeneous two-point boundary-value problems by an extension of Scott's version of invariant imbedding. More extensive computational results, including true systems (m,n>1) and comparisons of various alternatives (e.g. use of (14) over the entire interval) will be reported elsewhere.

Acknowledgment - This research was supported under Grant Number AFOSR 80-0235.

## References


1. M. R. Scott, Invariant Imbedding and its Applications to Ordinary Differential Equations, Addison-Wesley, Reading, Mass., 1973.
2. J. L. Casti and R. E. Kalaba, Imbedding Methods in Applied Mathematics, Addison-Wesley, Reading, Mass., 1973.
3. R. E. Bellman and G. M. Wing, An Introduction to Invariant Imbedding, John Wiley and Sons, New York, 1975.
4. D. O. Banks and G. J. Kurowski, Computations of Eigenvalues of Singular Sturm-Liouville Systems, Math. Comput. 22: 304-310 (1968).
5. I. T. Elder, An Invariant Imbedding Method for Computation of Eigenlengths of Singular Two-Point Boundary-Value Problems, J. Math. Anal. Appl., to appear.

6. W. R. Boland and P. Nelson, Critical Lengths by Numerical Integration of the Associated Riccati Equation to Singularity, Appl. Math. Comput. 1: 67-82 (1975).

7. P. Nelson and I. T. Elder, Calculation of Eigenfunctions in the Context of Integration-to-Blowup, SIAM J. Numer. Anal. 14: 124-136 (1977).

8. M. R. Scott and W. H. Vandevender, A Comparison of Several Invariant Imbedding Algorithms for the Solution of Two-Point Boundary-Value Problems, Appl. Math. Comput. 1: 187-218 (1975).

9. M. R. Scott, Invariant Imbedding and the Calculation of Internal Values, J. Math. Anal. Appl. 28: 112-119 (1969).

10. P. Nelson and M. R. Scott, Internal Values in Particle Transport by the Method of Invariant Imbedding, J. Math. Anal. Appl. 34: 628-643 (1971).

11. E. A. Coddington and N. Levinson, Theory of Ordinary Differential Equations, McGraw-Hill, New York, 1955.

12. L. F. Shampine and R. C. Allen, Numerical Computing, W. B. Saunders and Co., Philadelphia, 1973.

## DIFFICULTIES IN EVALUATING DIFFERENTIAL EQUATION SOFTWARE

by Robert D. Russell
Simon Fraser University
Burnaby, B.C. V5A 1S6/Canada

### I. Introduction

The recent proliferation mathematical software is accompanied by the responsibility for its evaluation. This has of course often been recognized by numerical analysts, and comparisons of many types have been conducted. In areas where the software is complex and multi-faceted, such as the numerical solution of boundary value problems for ordinary differential equations (BVODES), making a relative comparison of the performance of codes in order to recommend which is the most appropriate in a given situation is at best extremely difficult. In this paper, we shall discuss the types of comparisons which have been performed for BVODES, the limitations of these comparisons, and the problems which arise when these limitations are not fully appreciated. The current situation is well summarized in [1]: "Unfortunately the standards expected for mathematical exposition are only rarely applied to the reporting of computational experiments." While the focus here is on global methods for BVODES and their particular characteristics, the concerns and points expressed are equally applicable for most other software for differential equations and indeed many other areas of numerical analysis.

### II. History of BVODES

Comparison of BVODE methods, as for most other numerical methods, have traditionally been of 3 types:

(i) comparison of methods in a general way,
(ii) comparison of algorithms, and
(iii) comparison of codes.

In a comparison of type (i), it is assumed that methods can be linearly ordered, i.e. that it is reasonable to call one method better than another or others. For example, a recent numerical analysis book states unequivocally that multiple shooting is superior to finite differences and finite elements for solving difficult BVODES. In support of this statement, only one problem is solved, only equal spacing is used for the global (finite difference and finite element) methods, only low order global methods are used, and only one finite element method is considered. It is hopefully clear that comparisons of type (i) for popular, well-used methods will often border on the ridiculous because the objective is far too vague and unrealistic.

Comparisons of type (ii) have been quite popular, for it is natural to investigate which methods seem the most promising before investing the time to implement them in portable mathematical software. In [2] algorithms for the three common finite element methods (collocation, Rayleigh-Ritz-Galerkin, and least squares) are compared for an mth order linear BVODE, and on the basis of operation counts it is concluded that least squares is more efficient than collocation which in turn is more efficient than Rayleigh-Ritz-Galerkin. When a more efficient linear system solution scheme for collocation is considered the roles of least squares and collocation are switched [3]. Recently variants of these finite element methods have been compared (on the basis of CPU times for simple implementations), and collocation is concluded to be most efficient [4]. Obviously conclusions could be reversed if some more efficient versions are developed. Other examples of type (iii) comparisons are [5], and [6], where the first concludes that certain finite difference and initial value methods are more efficient than collocation methods, and the second, considering a more efficient implementation of collocation, concludes that it is competitive with finite differences.

Regardless of the conclusions drawn from a comparison of type (ii), it is still not clear that the type (iii) comparisons (i.e. of codes themselves) will give consistent results. Several components, such as the mesh selection and nonlinear iteration strategies, are difficult to evaluate adequately in terms of operation counts in algorithms, yet they are overriding features in determining the efficiency of the codes. This leads some to suggest that the only realistic definition of an algorithm for a numerical method is a computer code itself. Type (iii) comparisons are ultimately the most realistic ways to compare methods; unfortunately, difficulties still arise in areas such as BVODES from the complexity of the general programs. This is discussed further in future sections.

## III. Global BVODE Codes

The first BVODE codes were of the initial value type, incorporating the available robust initial value codes, usually with a shooting or multiple-shooting strategy (e.g., see [7], [8], [9]). The BVODE codes based upon global (non-initial value) methods which have been developed to the generality described below are PASVA3, a finite difference code using the trapezoidal rule with deferred corrections [10] and COLSYS, a spline collocation code [11].

Figure 1 shows how PASVA3 and COLSYS are more-or-less designed. The purpose of both programs is to produce an approximate solution accurate to within the user's requested error tolerance (TOL). Both also provide an estimate of the error (EST) in the global solution. The adaptive mesh selection and nonlinear iteration strategies are two features of the codes which have been extensively tested and enable them to solve fairly difficult problems. The mesh selection strategy operates as follows: after a solution has been computed on a mesh $\pi$, the error estimates are used to select a new mesh $\pi'$ such that hopefully the new approximate solution on $\pi'$ satisfies $|EST| \le TOL$ and the magnitude of the error is the same over each new subinterval (implying efficiency). Both nonlinear iteration strategies use a modified Newton method. The linear system for each iteration is solved using a Gauss elimination method which utilizes the special banded structure of the matrices.

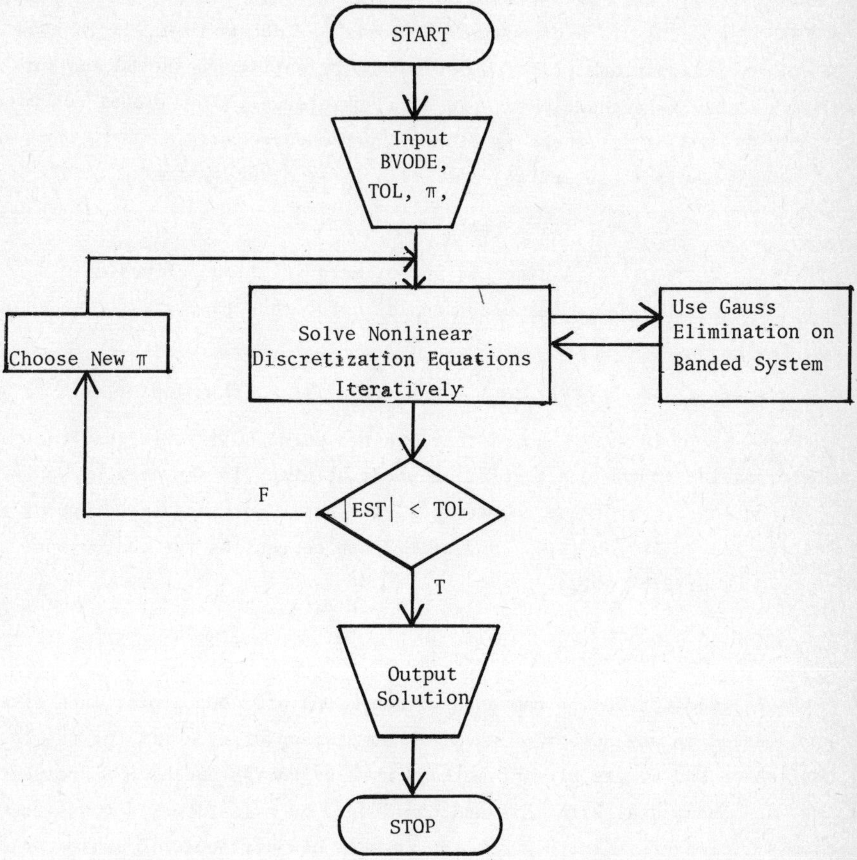


Figure 1

Even though the overall strategies for PASVA3 and COLSYS are similar, the underlying methods are quite different. Suppose a general BVODE is given, which consists of d nonlinear differential equations

$$u_i^{(m_i)}(x) = F_i(x, \underset{\sim}{z}(\underset{\sim}{u}(x))) \qquad a < x < b \qquad i = 1,\dots,d \tag{3.1}$$

where $\underset{\sim}{u}(x) = (u_1,\dots,u_d)^T$ is the sought solution vector and $\underset{\sim}{z}(\underset{\sim}{u}(x)) = (u_1, u_1', \dots, u_1^{(m_1-1)}, u_2, \dots, u_2^{(m_2-1)}, \dots, u_d^{(m_d-1)})^T$ is the vector of unknowns that would result from converting to a first order system, and $m^* = \sum_{i=1}^{d} m_i$ nonlinear, multi-point boundary conditions (BC). If the BC are separated COLSYS can solve the problem directly. Given a partition of [a,b]

(3.2) $\pi: a = x_1 < x_2 < \ldots < x_N < x_{N+1} = b \;,\; h = \max_{1\le j\le N} h_j = \max_{1\le j\le N}(x_{j+1}-x_j)$

and $k \ge \max_{1\le i\le d} m_i$ , the collocation approximation is a vector $\underset{\sim}{v} = (v_1,\ldots,v_d)^T$ such that $v_i \in C^{(m_i-1)}[a,b]$ is a piecewise polynomial of degree $k+m_i-1$ on each subinterval. The approximation $\underset{\sim}{v}(x)$ is determined by satisfying the BC and satisfying (3.1) at the $k$ Gauss-Legendre points in each subinterval. The global convergence rate is $O(h^{k+m})$, and on $\pi$ there is $O(h^{2k})$ superconvergence.

PASVA3 requires converting (3.1) to a first order system

(3.3) $$\underset{\sim}{z}'(x) = \underset{\sim}{F}(x;\underset{\sim}{z}) \qquad a \le x \le b$$

where $\underset{\sim}{z}(x) = (z_1,\ldots,z_{m^*})^T$ . A discrete approximate solution defined on $\pi$ , $\underset{\sim}{w}_j \doteq \underset{\sim}{z}(x_j)$ for $1 \le j \le N+1$ , is calculated by using the trapezoidal rule approximation to (3.3),

(3.4) $$\underset{\sim}{w}_{j+1} - \underset{\sim}{w}_j = \tfrac{1}{2}[F(x_j,\underset{\sim}{w}_j) + F(x_{j+1},\underset{\sim}{w}_{j+1})] \qquad 1 \le j \le N ,$$

and satisfying the BC (PASVA3 assumes they are two-point BC). This results in an $O(h^2)$ approximation to the exact solution whose accuracy is improved by applying deferred corrections. For these corrections, the basic mesh and structure of the linear systems remain as for (3.4), and after each correction the convergence is increased by $O(h)$ or $O(h^2)$ [12].

## IV. Difficulties with a Code Comparison

PASVA3 and COLSYS are composed of many separate components, each of which can be implemented in any number of ways. A comparison of type (ii) of the basic finite difference and finite element methods used by PASVA3 and COLSYS, respectively, would have to somehow deal with this and could only be reliable as a gross measure of relative efficiencies. In this section we show how difficult it is to compare the codes too, i.e. how difficult a comparison of type (iii) is.

In any evaluation of software, a first step is of course deciding what criteria to use. When codes are "relatively simple" and are written with consistent philosophies, as has been the case for many comparisons such as [13], [14], this can be straightforward. In the case of BVODE codes, however, selection of evaluation criteria which do not favor one code over another is virtually impossible. We shall somewhat arbitrarily split evaluation criteria into 3 groups.

The first group of criteria contains objective ones, and these normally include computer times and storage requirements, portability, and program correctness. Unfortunately, interpretation of results is not easy for even these objective criteria. For instance, CPU times for large codes or large systems of differential equations can vary significantly with the environment, depending for example upon local array handling facilities such as vector or parallel processors.

Since the mesh selection strategy of COLSYS can be affected by the amount of storage available, storage requirements for a given problem can be measured in different ways. While in a given environment we would not expect time or storage measurements to vary by orders of magnitude, a factor of 2-3 could easily be meaningless. Insuring portability and correctness of large programs demands time and resources often unavailable to individuals, so it is especially important in any comparison to carefully state what machines and compilers and what program parameter values are used to arrive at the numerical results. In [15] specific examples are given and these points are expanded upon.

The second group of evaluation criteria contains subjective ones, robustness and ease of use. These are in many ways the most difficult to measure and report on, but usually they are also among the most important. To make matters worse, they are in large measure opposing criteria in that modifying a program to make it more robust often makes it more difficult to use, and modification to facilitate its use often reduces the capabilities to handle difficult problems. As discussed below, COSLYS and PASVA3 have several options to improve robustness. Based upon the experience of many users of both codes, there seems to be little evidence that one is much more difficult to use than the other.

The third group of evaluation criteria arise from factors characteristic of the particular kind of codes being evaluated. For global BVODE codes, these factors include (a) form of the solution, (b) user feedback, (c) stopping criteria, (d) program parameters, (e) test problems, and (f) form of driver.

COLSYS provides the user with a piecewise polynomial solution expressed in terms of a B-spline basis, while PASVA3 provides a discrete solution defined at the mesh points. The degree of importance placed upon the availability of a global solution varies from user to user, and the way that a comparison deals with this issue cannot possibly be satisfactory from all perspectives. The same would generally be true of user feedback characteristics for BVODE codes. PASVA3 and COLSYS, however, have very similar options and capabilities. Unlike most codes based upon initial value type methods, they provide a regular record of how the mesh selection and nonlinear iteration are proceeding.

The codes stopping criteria are different, and this often results in their producing uncomparable results. PASVA3 attempts to have the absolute error for all of the approximate solution components less than TOL on $\pi$. For each approximation $v_r(x)$ to $u_r(x)$ in (3.1), COLSYS attempts to satisfy

$$\|u_r^{(\ell)} - v_r^{(\ell)}\|_i \le TOL_{r,\ell}(1+\|v_r^{(\ell)}\|_i) \qquad 1 \le i \le N \quad 0 \le \ell \le m_r-1$$

where $\|u\|_i = \max_{x_i \le x \le x_{i+1}} |u(x)|$ and $TOL_{r,\ell}$ $(0\le\ell\le m_r-1,\ 1\le r\le d)$ are user provided tolerances. When solution components and their derivatives vary significantly in magnitude, it is difficult to determine which tolerances cause the codes to provide similar accuracies [15]. Even if consistent stopping criteria have been determined, it is frequently necessary to perform complicated experimentation with the program parameters (such as an initial mesh and solution) to see which ones are preferable. For example, one code may perform best when a crude equal spaced mesh is input, and the other may be best when a fine unequal spaced initial mesh is used. Also, each code has its own special program parameters which can improve performance significantly. Decisions must somewhat arbitrarily be made as to how many parameters will be used, how much experimentation with different values will be done, and how many of the numerical results will ultimately be reported.

Problems of many types can be reformulated as "standard" BVODES acceptable by the codes, and these reformulations can frequently be done in several ways [16], [17]. This, together with the existence of a rich variety of kinds of BVODES, combine to make the task of selecting a representative set of test problems a most difficult one -- one which has until now not been satisfactorily completed.

Although of less importance, the form of the driver is not a consideration which can be neglected. Especially for large or highly sensitive BVODES, deciding which way to evaluate the coefficient functions and whether or not to use divided differences for evaluating Jacobians could be significant.

## V. Conclusions

Before a comparison of codes such as PASVA3 and COLSYS can be made, the difficulty of selecting reasonable evaluation criteria must be faced. Several efforts at comparing these codes have been discontinued due to inability to overcome this problem. Perhaps the most natural way of resolving it is to take a "greatest common denominator" approach -- to select the most flexible criteria which can be directly applied to both codes. For example, run the programs on the same BVODES (written in the same form), with the same tolerances, the same initial mesh and the same storage, measure the error at the same discrete set of points, and use continuation only if it is necessary. This is more-or-less the philosophy taken in the recent comparison [18]. Care is taken to see that the codes are treated as fairly as possible in this context, and the comparison is most interesting. Still, it is difficult to know how conclusions would be affected if the methodology is changed. For example, even if the same number of discretization equations are used initially (rather than subintervals), results could conceivably change dramatically on certain BVODES.

One problem with a comparison of complex codes such as COLSYS and PASVA3 is that results can easily be misinterpreted by unsophisticated users. For example, the ability of one code (and not the other) to solve a problem without continuation, even though it requires a large amount of CPU time, could be incorrectly considered a weakness unless its results with continuation were also reported alongside those for the other code. There are many other examples of possible dangers of misinterpretation. Thus, even assuming that reasonable evaluation criteria can be selected, there remains the problem of displaying results for straightforward interpretation.

In an evaluation of nonlinear least squares and optimization software, Hiebert [19], [20] displayed experimental results visually by use of Chernoff faces. The Chernoff faces in Figure 2 below, with the exception of the moustaches, appear in [19]. Each respresents the performance of one code on a set of problems, where the size of the face depends on the number of problems successfully solved, the width of the nose depends upon the number of returns because of too many function evaluations, etc.

Figure 2

Unfortunately, it is not obvious what criteria are reasonable for selecting the facial features and measurement scales for their displays. Moreover, the fact that undesirable subjective characteristics of the faces can enter into the evaluation, as the exaggerated example of the mustaches shows, questions further the validity of conclusions based upon these faces. An alternative could be the use of star plots (shown in Figure 3), where various criteria (11, in the figure) are selected and the measured performance of a code is displayed with a star-like diagram.

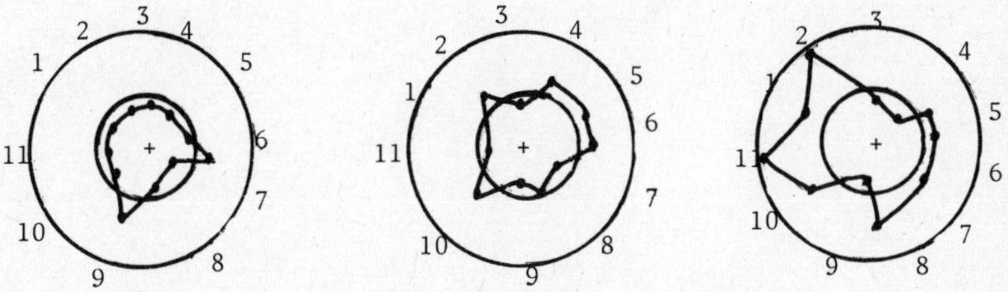


Figure 3

The considerations involved in software evaluation are by-and-large not very mathematical nor typical of numerical analysis research. Nevertheless, if numerical analysts are going to be involved in the development of mathematical software, they must be willing to share the responsibility for its testing and relative evaluation. The recent BVODE codes have enabled users to solve more efficiently (or even for the first time) difficult problems of many types, e.g. those arising from studies in coal gasification, seismic ray tracing, controlled fusion, elementary particle physics, shell buckling, nuclear reactions, singular perturbations theory, Hopf bifurcations, optimal control theory, controlled fusion, fluid flow, electron flow in transistors, catalytic convertors, and beam bending.

Even given the considerable difficulties of comparing codes, evaluations such as [18] are still necessary for making general recommendations to users about code characteristics. Despite their obvious flaws, they do initiate the development of improved experimental design tools and are generally preferable to the common situations where people select codes on the basis of the strength of the code designers' personalities.

VI. References


1. H. Crowder, R.S. Dembo, and J.M. Mulvey, On reporting computational experiments with mathematical software, ACM trans. on Math. Software 5 (1979), 193-203.
2. R.D. Russell and J.M. Varah, A comparison of global methods for linear two-point boundary value problems, Math. of Comp. 29 (1975), 1-13.
3. R.D. Russell, Efficiencies of B-spline methods for solving differential equations, Proc. 5th Manitoba Conference on Numerical Mathematics, Utilitas Math. (1975), 599-617.
4. J.C. Diaz, G. Fairweather, and P. Keast, A comparative study of finite element methods for the solution of second order linear two-point boundary value problems, Dept. of Computer Science Tech. Report 150/81, Univ. Toronto (1981).
5. H.B. Keller, Numerical solution of boundary value problems for ordinary differential equations: survey and some recent results on difference methods, Numerical Solutions of Boundary Value Problems for Ordinary Differential Equations, A.K. Aziz, ed., Academic Press, New York, 1975.
6. R.D. Russell, A comparison of collocation and finite differences for solving two-point boundary value problems, SIAM J. Numer. Anal. 14 (1977), 19-39.
7. R. Bulirsch, J. Stoer, and P. Deuflhard, Numerical solution of nonlinear two-point boundary value problems I, Numer. Math. Handbook Series Approximation (1976).
8. M.L. Scott and H.A. Watts, Computational solutions of linear two-point boundary value problems via orthonormalization, SIAM J. Numer. Anal. 14 (1977), 40-70.
9. I. Gladwell, Shooting codes in the NAG library, Codes for Boundary-Value Problems in Ordinary Differential Equations, B. Childs et al., ed., Springer-Verlag Lecture Notes in Computer Science 26 (1979).
10. M. Lentini and V. Pereyra, An adaptive finite difference solver for nonlinear two-point boundary value problems with mild boundary layers, SIAM J. Numer. Anal. 14 (1977), 91-111.
11. U. Ascher, J. Christiansen, and R.D. Russell, Collocation software for boundary value ODE's, ACM Trans. on Math. Software (1981).
12. J. Christiansen and R.D. Russell, Deferred corrections using uncentered end formulas, Num. Math. 35 (1980), 21-33.
13. D. Kahaner, Comparison of numerical quadrature formulas, in Mathematical Software, ed. by J. Rice, Academic Press, New York (1971), 229-259.
14. L.F. Shampine, H.A. Watts, and S.M. Davenport, Solving nonstiff ordinary differential equations -- the state of the art, SIAM Review 18 (1976), 376-411.
15. V. Pereyra and R.D. Russell, Difficulties of comparing complex mathematical software: general comments and the BVODE case, manuscript (1981).

16. V. Pereyra, Solución numérica de ecuaciones diferenciales con valores de frontera, Acta Cient. Venezolana 30 (1979), 7-22.

17. U. Ascher and R.D. Russell, Reformulation of boundary value problems into "standard" form, SIAM Review 23 (1981), 238-254.

18. P.W. Hemker, H. Schippers, and P.M. de Zeeuw, Comparing some aspects of two codes for two-point boundary-value problems, Math. Centrum Report NW 98/80, Amsterdam (1980).

19. K.L. Hiebert, A comparison of software which solves systems of nonlinear equations, Sandia Labs Report SAND 79-0483, Albuquerque, N.M. (1979).

20. K.L. Hiebert, A comparison of nonlinear least squares software, Sandia Labs Report 80-0181, Albuquerque, N.M. (1980).

TOPICS IN FINITE ELEMENT DISCRETIZATION
OF PARABOLIC EVOLUTION PROBLEMS

J.P. HENNART
Instituto de Investigaciones en Matemáticas
Aplicadas y en Sistemas
Universidad Nacional Autónoma de México
Apartado Postal 20-726, México 20, D.F.

## 1. INTRODUCTION

This paper is devoted to some practical implementation aspects of the finite element method for parabolic evolution problems. In its first part, it is shown how reduced integration techniques [1, Chapter 20] used in conjunction with tensor product rectangular elements eventually lead to an interesting alternative to the Galerkin ADI schemes proposed by Douglas and Dupont [2].

In the second part, time discretization aspects are considered and it is shown how a class of finite elements in time schemes which has been presented before [3] can be quite efficiently implemented using the iterated defect correction idea developed in [4] for collocation methods applied to stiff ODE's

## 2. SPACE DISCRETIZATION ASPECTS

The application of standard semidiscrete Galerkin techniques [5] to mixed initial-boundary value parabolic evolution problems typically leads to (large) systems of stiff ODE's of the general form

$$M\,\dot{\underline{U}} + K(t,\underline{U})\underline{U} = \underline{F}(t) \qquad (2.1)$$

subject to appropriate initial conditions

$$\underline{U}(0) = \underline{U}_0 \quad . \qquad (2.2)$$

If $S^k \equiv \text{span}\{u_i,\ i=1,\dots,N\} \subset H^1_0(\Omega)$ is the finite element subspace used, M and K are the mass and stiffness matrices respectively and

$$M = (m_{ij}) \quad , \quad \text{with } m_{ij} = (u_i,u_j) \quad , \qquad (2.3)$$

while

$$K = (k_{ij}) \quad , \quad \text{with } k_{ij} = a(u_i,u_j) \quad , \qquad (2.4)$$

where

$$a(u,v) = \int_\Omega p\nabla u^T . \nabla v \; d\overline{r} \qquad , \qquad (2.5)$$

if the elliptic operator involved is taken to be $\nabla . p \nabla$ for the sake of simplicity. In (2.5), p depends on the space and time variables and eventually on the solution u in nonlinear situations. Over tensor product domains such as the rectangle $[a,b] \times [c,d]$ in $\mathbb{R}^2$, a fairly natural choice consists in taking basis functions $u_i$ of the tensor product type, namely

$$S^k \equiv \{\alpha_1(x)\beta_1(y), \ldots, \alpha_{N_x}(x)\beta_{N_y}(y)\} \qquad , \qquad (2.6)$$

so that following Douglas and Dupont [2],

$$M = M_x \otimes M_y \qquad , \qquad (2.7)$$

and if $p=\overline{p}$ constant,

$$K = \overline{K} = \overline{p}(K_x \otimes M_y + M_x \otimes K_y) \qquad , \qquad (2.8)$$

where

$$M_x = (m_{x,ij}) \quad , \quad \text{with } m_{x,ij} = \int_a^b \alpha_i(x)\alpha_j(x)dx \; , \qquad (2.9)$$

$$M_y = (m_{y,ij}) \quad , \quad \text{with } m_{y,ij} = \int_c^d \beta_i(y)\beta_j(y)dy \; , \qquad (2.10)$$

$$K_x = (k_{x,ij}) \quad , \quad \text{with } k_{x,ij} = \int_a^b \alpha_i'(x)\alpha_j'(x)dx \; , \qquad (2.11)$$

etc., $\otimes$ in (2.7) and (2.8) denoting the standard matrix tensor product.

With $A \equiv -M^{-1}K$ and assuming that $\underline{F}=\underline{0}$ in (2.1) again for the sake of simplicity, (2.1) can be rewritten as

$$\dot{\underline{U}} = A\underline{U} \qquad , \qquad (2.12)$$

where

$$A = \overline{A} = \overline{A}_x \otimes I_y + I_x \otimes \overline{A}_y \qquad , \qquad (2.13)$$

with

$$\overline{A}_x \equiv -\overline{p}M_x^{-1}K_x \quad \text{and} \quad \overline{A}_y \equiv -\overline{p}M_y^{-1}K_y \quad , \tag{2.14}$$

$I_x$ and $I_y$ being the identity matrices of order $N_x$ and $N_y$ respectively.

In (2.12), A can still depend on t and on $\underline{U}$ in the nonlinear case. If we accept to be limited to second order (global) accuracy in time, A can always be averaged over the time step of interest and nonlinearity can be eliminated by using either a predictor-corrector or an extrapolation scheme as in Ref. 5. In any case, A becomes a constant in time matrix and the formal solution of (2.12) over $[t_i, t_{i+1}]$ is

$$\underline{U}(t_{i+1}) = \exp(h\overline{A})\underline{U}(t_i) \quad , \tag{2.15}$$

or, since $\overline{A}_x \otimes I_y$ and $I_x \otimes \overline{A}_y$ commute,

$$\underline{U}(t_{i+1}) = \exp(h\overline{A}_x \otimes I_y) . \exp(hI_x \otimes \overline{A}_y)\underline{U}(t_i) \quad , \tag{2.16}$$

which can be approximated by

$$\underline{U}_{i+1} = [I - \Theta_x h\overline{A}_x \otimes I_y]^{-1} [I + (1-\Theta_x)h\overline{A}_x \otimes I_y]\underline{U}_{i+1/2} \quad ,$$

$$\underline{U}_{i+1/2} = [I - \Theta_y hI_x \otimes \overline{A}_y]^{-1} [I + (1-\Theta_y)hI_x \otimes \overline{A}_y]\underline{U}_i \quad . \tag{2.17}$$

In (2.17), $\Theta_x$ and $\Theta_y$ are functions of the time step h such that the rational approximations to the exponentials of (2.16) are of second order (globally) and therefore consistent with the (global) second order accuracy mentioned hereabove. For more details, see [6-7]. With an appropriate ordering of the nodes, the two basic steps in (2.17) boil down to solving $N_x$ one-dimensional y problems ($\underline{U}_i \rightarrow \underline{U}_{i+1/2}$) and then $N_y$ one-dimensional x problems ($\underline{U}_{i+1/2} \rightarrow \underline{U}_{i+1}$).

In the general case where p depends on $\overline{r}$, $A \neq \overline{A}$ and the above approach is not applicable. In [2], Douglas and Dupont essentially proposed to replace A in (2.12) by $\overline{A} + (A-\overline{A})$, where $\overline{A}$ is obtained from some mean value $\overline{p}$ of p. As a result, Eq. (2.12) becomes

$$\dot{\underline{U}} = \{\overline{A} + (A-\overline{A})\}\underline{U} \quad . \tag{2.18}$$

so that

$$\underline{U}(t_{i+1}) = \exp[h\{\bar{\bar{A}} + (A-\bar{\bar{A}})\}]\underline{U}(t_i)$$

$$= \exp(h\bar{A}).\exp\{h(A-\bar{A})\}.U(t_i)+O(h^2) \quad , \quad (2.19)$$

since $\bar{A}$ and $(A-\bar{A})$ do not commute. Using an implicit Backward Euler (BE) scheme for $\exp(h\bar{A})$ and an explicit Forward Euler (FE) scheme for $\{h(A-\bar{A})\}$ leads after reordering to the following final scheme

$$[I-h\bar{A}_x \otimes I_y]\,[I-hI_x \otimes \bar{A}_y]\,(\underline{U}_{i+1}-\underline{U}_i)$$

$$= hA\underline{U}_i + O(h^2) \quad , \quad (2.20)$$

which exhibits the same decoupling features as (2.17). All the time variations of A only affect the second member and this leads to interest time stepping strategies where the left handside matrices are rebuilt and factorized from time to time only, depending on the time variations of A and on the conditions that they induce on $\bar{\bar{A}}$ to achieve unconditional stability. For more details on this reinterpretation of the Galerkin ADI schemes of Ref. 2, see [8] where some extensions are also proposed.

The more direct approach to decoupling that we propose here is based on the use of reduced integration techniques [1, Chapter 20]. Let us write K as $K^x+K^y$ where $K^x$(resp. $K^y$) is that part of the stiffness matrix K which corresponds to the derivatives in the x (resp. y) direction. In standard Galerkin implementations, too much coupling results as the operator $\partial_x p \partial_x$ couples the unknowns not only in the x direction but also in the y direction, leading for instance to a 9 to 25 points coupling with biquadratic elements (see Figure 1).

Let us consider a matrix element $k^x_{ij}$ of $K^x$ evaluated by numerical integration

$$k^x_{ij} = \int_\Omega p u_{ix}\, u_{jx}\, d\bar{r}$$

$$= \int_\Omega p(\alpha'_k\, \beta_\ell)\,(\alpha'_m\, \beta_n)\, dxdy$$

$$\sim \sum_q w_q p_q (\alpha'_{kq}\, \beta_{\ell q})\,(\alpha'_{mq}\, \beta_{nq}) \quad , \quad (2.21)$$

where $\quad p_q \equiv p(\overline{r}_q), \quad \alpha'_{kq} \equiv \alpha'_k(x_q)$ and $\beta_{\ell q} \equiv \beta_\ell(y_q)$.

Clearly if we choose the sampling points $\overline{r}_q=(x_q,y_q)$ to be the nodes of the tensor product element considered, $\beta_{\ell q}=\delta_{\ell q}$ and $\beta_{nq}=\delta_{nq}$ so that coupling through $K^x$ only exists if $q=\ell=n$, in other words between those nodes which have the same y coordinate. In contrast with the standard implementation, this now leads to a 3 to 5 points coupling with the same biquadratic element considered before (see Figure 2).

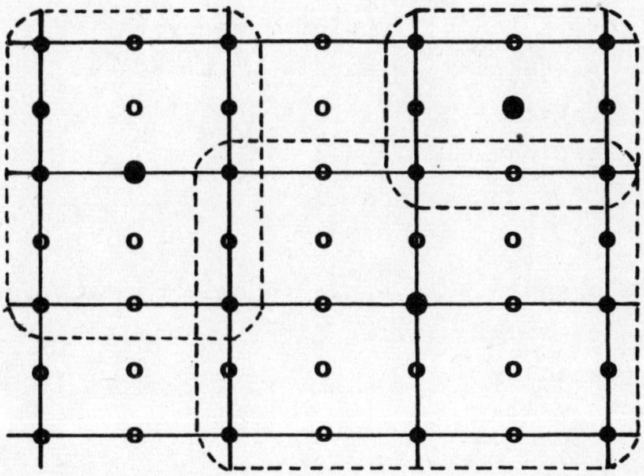

Figure 1. 9 to 25 points couplings for biquadratic elements with standard Galerkin semidiscretization ($\partial_x p \partial_x o$ or $\partial_y p \partial_y o$)

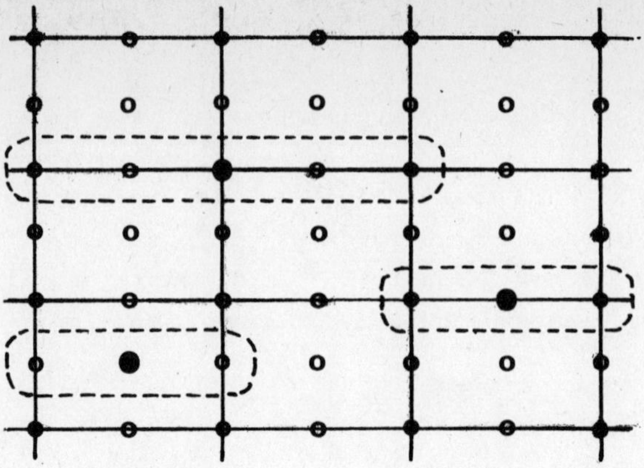

Figure 2. 3 to 5 points couplings for biquadratic elements with reduced integration Galerkin semidiscretization $(\partial_x p \partial_x o)$

Using the same arguments as above to eliminate the explicit time dependence and the eventual nonlinearity, we still have Eq. (2.12) where A is now replaced by

$$\overline{A} = \overline{K}^x + \overline{K}^y \quad , \tag{2.22}$$

the bars denoting the time averaged and (or) linearized forms consistent with second order (global) accuracy. The form (2.22) is valid for any coefficient p, and does not depend as in (2.13) on its constancy (or at least its factorizability into functions of one coordinate only). Since $\overline{K}^x$ and $\overline{K}^y$ do not necessarily commute, Eq. (2.15) leads to

$$\underline{U}(t_{i+1}) = \exp(h\overline{K}^x).\exp(h\overline{K}^y)\underline{U}(t_i) + O(h^2) \quad , \tag{2.23}$$

or better to

$$\underline{U}(t_{i+1}) = \exp(h\overline{K}^x/2).\exp(h\overline{K}^y).\exp(h\overline{K}^x/2)\underline{U}(t_i) + O(h^3). \tag{2.24}$$

Because of the couplings induced by $\bar{K}^x$ and $\bar{K}^y$ after reduced integration, any rational approximation to these exponentials as in (2.17) would lead to sequences of one-dimensional problems.

In [9], an experimental space-time kinetics code has been developed and applied to nuclear reactor modelling in two dimensions. The spatial approximation was treated with the biquadratic finite elements used as examples hereabove. For time discretization, different ADI and LOD approximations based on Eq. (2.23) or (2.24) were used (see e.g. Ref. 10), as well as ADI-$B^2$ and LOD-$B^2$ schemes developed with the same ideas as in [11]. Preliminar numerical results reported in [9] show that the techniques proposed are quite efficient at least for the simple test problems considered up to now. More extended comparisons with fairly complex benchmark problems will be reported later [12].

## 3. TIME DISCRETIZATION ASPECTS

For a given partition $\pi \equiv \{t_i, i=0,\ldots,I\}$ of $[0,T]$, integration of (2.1) over $[t_i, t_{i+1}]$ gives

$$M\underline{U}(t_{i+1}) = M\underline{U}(t_i) - \int_{t_i}^{t_{i+1}} (K(t)\underline{U}(t) - \underline{F}(t))dt, \tag{3.1}$$

where it is assumed for simplicity that K does not depend on $\underline{U}$. Eq. (3.1) can then be approximated by

$$M\underline{U}_{i+1} = M\underline{U}_i - \int_{t_i}^{t_{i+1}} (K(t)\underline{U}_{pq}(t) - \underline{F}(t))dt, \tag{3.2}$$

where p,q are nonnegative integers with $p+q=\ell+1 > 0$, each component of $\underline{U}_{pq}$ belonging to a linear space $S^h$ of dimension p+q, with in general polynomial and (or) exponential basis functions, a natural choice for parabolic evolution problems. In $S^h$, a generalized Hermite interpolant is introduced with p(resp. q) interpolation conditions at $t=t_i$ (resp. $t_{i+1}$). By collocation at $t=t_i$ and (or) $t=t_{i+1}$, the successive derivatives can be eliminated and the only parameters left turn out to be $\underline{U}_i^{(0)}=\underline{U}_i$, known from the initial conditions or from previous time integration, and $\underline{U}_{i+1}^{(0)}=\underline{U}_{i+1}$. The result is a class of one-step integration methods, each particular member of which depending on the quadrature scheme (QS) used to approximate the righthand side of (3.2). For more details, see [3]. With nonlinear ODE's, the case where $S^h \equiv P_\ell$, the linear space of polynomials of degree less than or equal to $\ell$, has been treated in [13]. Extensions to exponential bases leading to exponentially fitted schemes were introduced in [14] and studied in [6]. Generalizations with collocation at intermediate points (and not only at $t_i$ and $t_{i+1}$) are proposed in [7].

Since the (global) order in time of these finite element oriented schemes is ($\ell$+1) and since they are A-stable if q=p and in particular L-stable if q=p+1 or p+2 (at least in the polynomial case), it was pointed out in a companion paper [15] that among the second order schemes were schemes with good asymptotic properties (L-stability) but of first order only, like the BE scheme, or schemes of second order but with bad asymptotic properties leading to oscillations like the Crank-Nicolson (CN) one. It seems therefore that among the best candidates among the class of schemes proposed are p=1, q=2 schemes which are of third order in time. Two bases come to the mind:

$$S_o^h \equiv \{1,t,t^2\} \quad , \qquad (3.3)$$

or

$$S_\lambda^h \equiv \{1,t,\exp(\lambda t)\} \quad . \qquad (3.4)$$

Both lead to schemes with good asymptotic behavior if $\lambda$ is real negative. As shown in [6,13], the QS used must be exact for members of $S^h$. For $S_o^h$, let us mention for instance the Lobatto three-points QS (L3) and the Radau two-points righthand QS (R2) with the following weights $w_j$ and abscissae $t_{ij}=t_i+\tau_j h$

$$\{w_j\} = \{h/6,\ 2h/3,\ h/6\} \quad ,$$

$$\{\tau_j\} = \{\ 0\ ,\ 1/2,\ 1\ \} \quad , \qquad (3.5)$$

(L3 QS)

and

$$\{w_j\} = \{\ 3h/4,\ h/4\ \} \quad ,$$

$$\{\tau_j\} = \{\ 1/3\ ,\ 1\ \} \quad . \qquad (3.6)$$

(R2 QS)

Explicit expressions for the L3 and R2 (1,2) schemes with $S_o^h$ have been given in [15]: the R2 scheme is certainly to be preferred, in particular because it only requires that K be evaluated at two intermediate times. The practical implementation of both schemes however present some obvious difficulties, all of which have to do with the presence of a matrix $K_\alpha M^{-1} K_\beta$ in the lefthand side of the system of algebraic equations to be solved at each time step. A fairly complete discussion of these difficulties as well as of some of the remedies which have been proposed up to now to cope with them can be found in [15].

If instead of $S_0^h$, $S_\lambda^h$ is used, the main difference stems from the fact that the QS to be used must now be exact for $S_\lambda^h$. If as above we prefer the R2 scheme, for each given $\lambda$ and $h$ we have to solve a small system of nonlinear equations in $w_1$, $w_2$ and $\tau_1$ $(\tau_2=1)$, namely

$$w_1 + w_2 = 1 \quad ,$$

$$w_1\tau_1 + w_2 = 1/2 \quad ,$$

and

$$(\exp(\lambda h)-1)/\lambda h = w_1 \exp(\lambda h\tau_1) + w_2 \exp(\lambda h) \quad . \qquad (3.7)$$

In practice, since the weight and abscissae (3.6) are fairly good first guesses, the numerical solution of (3.7), for instance by some Newton-Raphson iteration, is straightforward and relatively unexpensive. Our numerical experiments have shown that less than 10 iterations are needed most of the time, which is quite negligible with respect to the simple cost of moving one step forward.

The practical implementation we are proposing for the above schemes essentially copes with all the difficulties mentioned hereabove. It is based on the iterated defect correction idea developed in [4,16], to which we refer the reader for more details. In [4], Frank and Heberhuber applied that idea to ordinary collocation techniques on uniform meshes. After realizing that the R2 scheme is an ordinary collocation scheme which collocates at $\tau_1$ and $\tau_2=1$ (by definition), or that the L3 scheme is a generalized collocation scheme (in the sense of [16]) which collocates at $\tau_1=0$ and $\tau_2$ in a weighted way as well as at $\tau_3=1$ (by definition), it is easy to generalize the formalism originally proposed in [4] to such situations.

The defect correction idea applied to the R2 schemes implies the following steps on each time interval:

1/ Solve (2.1) from $t_i$ to $t_{i1}=t_i+\tau_1 h$ and from $t_{i1}$ to $t_{i+1}$ using the BE scheme, namely

$$[M + \tau_1 h K_{i1}]\underline{U}^0_{i1} = M\underline{U}^0_i + \tau_1 h \underline{F}_{i1} \quad , \qquad (3.8a)$$

$$[M + (1-\tau_1)h K_{i+1}]\underline{U}^0_{i+1} = M\underline{U}^0_{i1} + (1-\tau_1)h \underline{F}_{i+1} \quad , \qquad (3.8b)$$

where $\underline{U}^0_i \equiv \underline{U}_i$. The result of this first step is $\underline{U}^0 = [\underline{U}^0_{i1}, \underline{U}^0_{i+1}]$.

2/ Build up the vector $\underline{P}^0(t)$ of components belonging to $S^h_0$ or $S^h_\lambda$ interpolating $\underline{U}^0_i$, $\underline{U}^0_{i1}$ and $\underline{U}^0_{i+1}$, and solve

$$M\dot{\underline{V}}^0 \equiv -K\underline{V}^0 + M\dot{\underline{P}}^0 + K\underline{P}^0 = -K\underline{V}^0 + \underline{F}^0 \quad , \qquad (3.9)$$

with the same strategy as in 1/ and the same initial conditions. The result is $\underline{V}^0 = [\underline{V}^0_{i1}, \underline{V}^0_{i+1}]$. Because problem (3.9) is a neighboring problem with a known solution ($\underline{P}^0$) to the initial problem (2.1), the exact error $\underline{V}^0 - \underline{P}^0$ at the nodes $t_{i1}$ and $t_{i+1}$ can be assumed to be a fairly good approximation to the same error for problem (2.1). $\underline{U}^0$ can be corrected accordingly

$$\underline{U}^1 = \underline{U}^0 - (\underline{V}^0 - \underline{P}^0) \quad , \qquad (3.10)$$

this last step being eventually repeated in an iterative way

$$\underline{U}^{n+1} = \underline{U}^0 - (\underline{V}^n - \underline{P}^n) \quad , \qquad (3.11)$$

where $\underline{P}^n$ has components in $S^h_0$ or $S^h_\lambda$ interpolating the corresponding components of $\underline{U}^0_i$, $\underline{U}^n_{i1}$ and $\underline{U}^n_{i+1}$. In this "local iteration strategy", the approximation is improved possibly until convergence before we move to the next time interval. From (3.11) a fixed point will eventually be reached if $\underline{U}^{n+1} = \underline{U}^n = \underline{P}^n$, that is if $\underline{U}^0 = \underline{V}^n$, implying from (3.9) that

$$\underline{F}^n = M\dot{\underline{P}}^n + K\underline{P}^n = \underline{F} \quad \text{at} \quad t = t_{i1} \text{ and } t_{i+1} \quad . \qquad (3.12)$$

In other words, $\underline{P}^n$ *turns out to be the solution of our R2 collocation scheme.*

In general, the iterative procedure (3.11) can be rewritten as

$$\underline{U}^{n+1} = S\,\underline{U}^n + \underline{G} \quad , \qquad (3.13)$$

and the existence of a fixed point requires that the spectral radius $\rho(S)$ of the iteration matrix S be less than one

$$\rho(S) < 1 \quad . \tag{3.14}$$

Applying the above formalism to the test equation $\dot{u}=\mu u, \mu \in \mathbb{R}$, it is easy to obtain the spectral radius of S as a function of $z=\mu h$ for different values of $\lambda$ and in particular for $\lambda=0$. As shown in Fig. 3, $\rho$ exhibits a quite favorable behavior for different values of $\lambda$ of interest and the iterative procedure (3.11) should be rapidly convergent, as we indeed verified. In Refs. 17 and 18, there is a lenghty discussion about the fact that for iterated defect schemes with nonequidistant nodes as hereabove, the successive iterates may remain of the same order of accuracy until we get in the vicinity of the fixed point where a rapid but unsystematic rise of the order occurs. For small systems of stiff ODE's, this may eventually lead to prefer collocation schemes with equidistant nodes. For parabolic PDE's where the extra amount of computation associated with one iteration may be quite negligible as we shall see, we feel that the existence in itself of a fixed point is the main consideration independently of the convergence pattern to this fixed point.

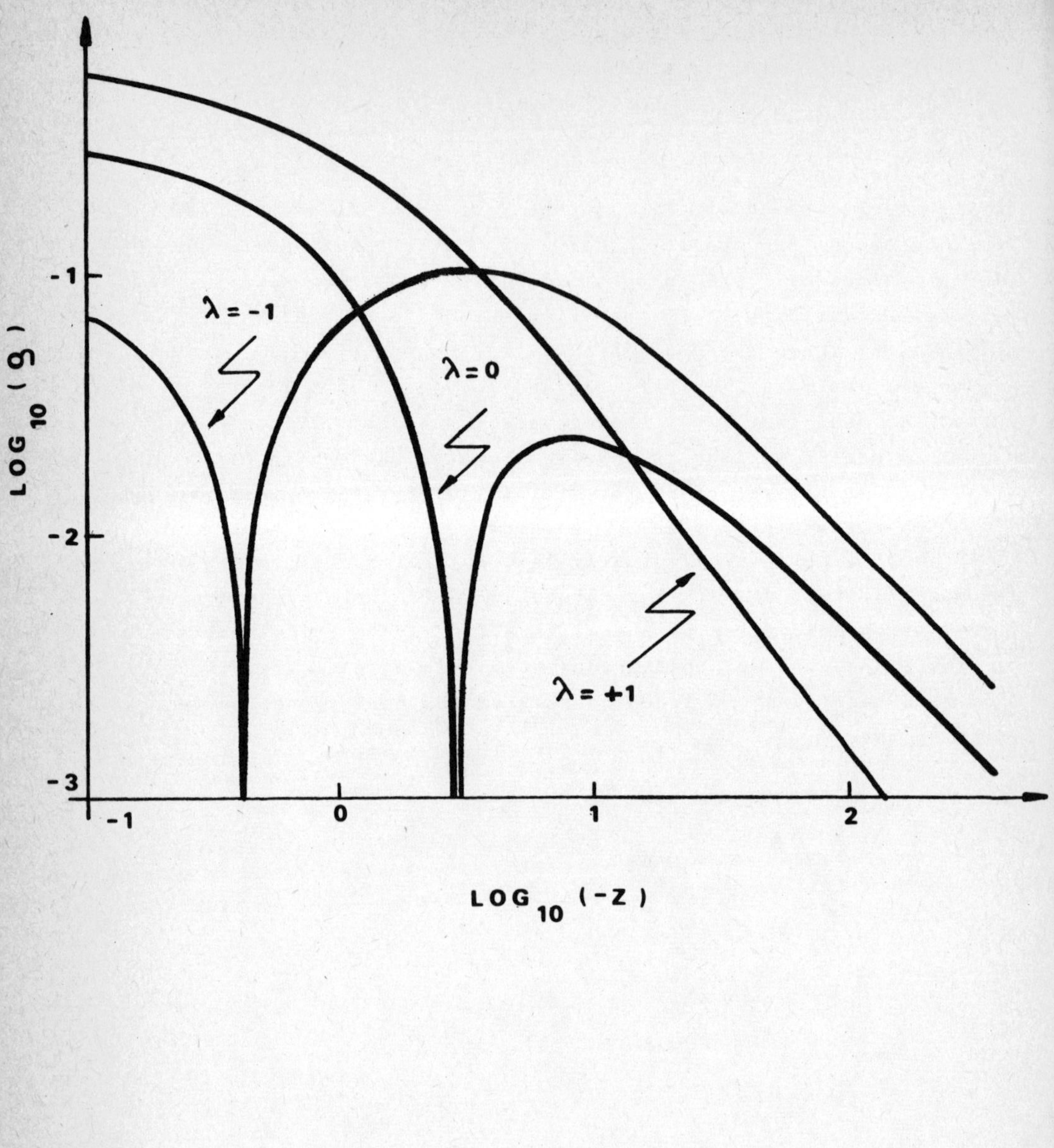


Figure 3. $\rho$(S) versus (-z)

A simple 1D parabolic sample problem has been solved numerically, namely

$$u_t = pu_{xx} - qu \quad , \quad (x,t) \in [0,1] \times [0,0.1] \quad , \tag{3.15}$$

with time-dependent coefficients p and q chosen such that the solution is

$$u(x,t) = \sin\pi x . \exp(-\pi^2 t) + \sin 9\pi x . \exp(-81\pi^2 t).(1+t). \tag{3.16}$$

On our Burroughs B6800 computer, a direct implementation of the R2 scheme with $S_0^h$ over a 20x20 space-time mesh required 275 secs. of CPU time. Essentially the same result (up to 7 significant decimal digits) was achieved in 28 secs. using iterated defect correction in an adaptive way with $S_0^h$. Even better results were obtained with $S_\lambda^h$ where $\lambda$ was taken to be $(-\pi^2)$ as shown in Table I, and essentially with the same computing times.

TABLE I

Discrete $L_2$ norm of the error at t=0.1 sec (1D same problem)

| Space-time mesh | R2 scheme with $S_0^h$ | R2 scheme with $S_\lambda^h$ |
|---|---|---|
| 10 x 10 | 6.5(-6) | 3.3(-6) |
| 20 x 20 | 5.9(-7) | 1.7(-7) |

It is important to realize that solving (3.9) repeatedly with the same strategy as in (3.8) only implies second members manipulations for linear algebraic systems, whose matrices have been built up and factorized once for all in step 1 (Eqs. 3.8). Accordingly, the above figures are only indicative since for 1D problems the solution time is proportional to the factorization time so that each correction almost costs as much as the basic BE steps. For multidimensional problems, it should be clear that the correction phase is almost unexpensive, compared with the matrix factorizations associated with step 1. At the same time, the direct implementation of the R2 scheme becomes prohibitively expensive and all reasonable implementations must be based on elementary steps where only matrices of the form $M+\alpha hK_\beta$ are built up and factorized, as provided for by standard existing software.

A final comment seems to be of order: the proposed implementation could conveniently be combined with the one proposed in [19] where the emphasis is on trying to preserve the same high order accuracies while moving ahead in time with a scheme similar to the ones proposed in [6, 20] but *constant in time*. In a sense, we are showing in this paper how the basic high order scheme in [19] can be efficiently implemented.

## REFERENCES


1. O.C. ZIENKIEWICZ, The Finite Element Method, 3rd Edition. McGraw Hill, London (1977).

2. J. DOUGLAS, Jr. and T. DUPONT, Alternating-Direction Galerkin Methods on Rectangles. pp. 113-214 of B. Hubbard (Ed.), SYNSPADE 1970. Academic Press, New York (1971).

3. J.P. HENNART, Multiple Collocation Finite Elements in Time for Parabolic Evolution Problems. pp. 271-278 of J.R. Whiteman (Ed.), The Mathematics of Finite Elements and Applications III. Academic Press, London (1979).

4. R. FRANK and C.W. UEBERHUBER, Iterated Defect Correction for the Efficient Solution of Stiff Systems of Ordinary Differential Equations. BIT 17, 146-159 (1977).

5. J. DOUGLAS, Jr. and T. DUPONT, Galerkin Methods for Parabolic Equations. SIAM J. Numer. Anal. 7, 575-626 (1970).

6. J.P. HENNART and H. GOURGEON, One-Step Exponentially Fitted Piecewise Continuous Methods for Initial Value Problems. Comunicaciones Técnicas, Serie NA, No. 263, 43 p. (1981).

7. H. GOURGEON and J.P. HENNART, A Class of Exponentially Fitted Piecewise Continuous Methods for Initial Value Problems. In these Proceedings.

8. J.P. HENNART, Some Recent Numerical Methods for the Solution of Parabolic Evolution Problems. Séminaire d'Analyse Numérique No. 298, 27 p., Université Scientifique et Médicale de Grenoble, Mathématiques Appliquées (1978).

9. E. DEL VALLE G., Aplicación del Método de Elementos Finitos a la Dinámica de Reactores Nucleares. Tesis de Maestría en Ingeniería Nuclear, 169 p., Instituto Politécnico Nacional, México, D.F. (1981).

10. A.R. GOURLAY, Splitting Methods for Time Dependent Partial Differential Equations. pp.757-795 of D.A.H. Jacobs (Ed.), The State of the Art in Numerical Analysis. Academic Press, London (1977).

11. L.A. HAGEMAN and J.B. YASINSKY, Comparison of Alternating-Direction Time Differencing Methods with other Implicit Methods for the Solution of the Neutron Group-Diffusion Equations. Nucl. Sci. Engng. 38, 8-32 (1969).

12. J.P. HENNART and E. DEL VALLE G., In Preparation.

13. J.P. HENNART, One-Step Piecewise Polynomial Multiple Collocation Methods for Initial Value Problems. Math. Comp. 31, 24-36 (1977).

14. J.P. HENNART, Some Recent One-Step Formulae for Integrating Stiff Systems of ODE's. Université Scientifique et Médicale de Grenoble, Mathématiques Appliquées, Séminaire d'Analyse Numérique No. 253, 17/06/76, 31 p. (1976).

15. J.P. HENNART, On the Implementation of Finite Elements in Parabolic Evolution Problems. To Appear in J.R. Whiteman (Ed.), The Mathematics of Finite Elements and Applications IV. Academic Press, London.

16. C.W. UEBERHUBER, Implementation of Defect Correction Methods for Stiff Differential Equations. Computing 23, 205-232 (1979).

17. R. FRANK, J. THIEMER and C.W. UEBERHUBER, Numerische Quadratur mittels Iterierter Defektkorrektur. Report No. 34/78. Inst. f. Num. Math., Technical University of Vienna (1978).

18. E. HAIRER, On the Order of Iterated Defect Correction. Numer. Math. 29, 409-424 (1978).

19. J.H. BRAMBLE and P.H. SAMMON, Efficient Higher Order Single Step Methods for Parabolic Problems. Part I. Math. Comp. 35, 655-677 (1980).

20. J. DESCLOUX and N.R. NASSIF, Padé Approximations to Initial Value Problems. Report EPFL, Lausanne (1975).

# A CLASS OF EXPONENTIALLY FITTED PIECEWISE CONTINUOUS METHODS FOR INITIAL VALUE PROBLEMS

H. GOURGEON and J.P. HENNART*

*Instituto de Investigacions en Matematicas Aplicadas y en Sistemas*
*Universidad Nacional Autonoma de México*
*Apartado Postal 20-726, México 20, D.F.*

* *Present address : Centre de Mathématiques Appliquées, Ecole Polytechnique, 91128 Palaiseau cédex (France)*

## 1 - INTRODUCTION

The approximate solution of parabolic evolution problems is obtained in a fairly standard way by Galerkin semidiscretization of the space variables followed by some finite differencing in time [1].

The space discretization by finite elements usually leads to high order approximations in space, especially for parabolic evolution problems which exhibit exponential smoothing in time. Consistent high order approximations in time result from the use of piecewise continuous approximations in time also, for instance piecewise polynomial ones [2].

For evolution equations of the parabolic type, a fairly natural idea consists of using not only polynomials but also exponentials in time. This leads to the exponential fitting concept [3] by which some predominant decay modes of the physical system at hand can hopefully be followed with full numerical accuracy.

The basic formalism using piecewise polynomials and exponentials in time for parabolic evolution problems was proposed in [4]. In [5,6], corresponding practical implementation details are discussed. In this paper, we shall restrict ourselves to the case of a (small) system of nonlinear stiff ordinary differential equations with specified initial conditions and show how exponential fitting can be provided in a finite element context while retaining the optimal convergence orders obtained in [2]. In a sense, this paper is closely related to a companion paper [7], where all the basic theorems were proved in an "ad hoc" manner. Here extensive use is made of the Tchebycheffian systems theory following the particularly lucid account of them which was recently given in [8]. This leads in particular to fairly obvious extensions of the formalism given in [7]. These extensions, as well as extensions to fully parabolic problems, will be detailed in forthcoming papers. The main theorems are given in Section 2, while some brief examples are proposed in Section 3.

## 2 - GENERAL MULTIPLE COLLOCATION FORMALISM

Let us consider the following nonlinear first-order differential equation

$$Dy = f(y,t) , \qquad t \in [0, T] , \tag{2.1}$$

subject to the initial condition

$$y(0) = y_o . \tag{2.2}$$

For the sake of simplicity, one equation only is considered, although all the results of this paper can be extended in a straight forward way to the case of a system of equations.

Let $\Pi : t_i,\ i = 0, \ldots, N$, with $t_o = 0$, $t_N = T$ and $h = \max_i h_i$ where $h_i = t_{i+1} - t_i$, be some mesh over $[0,T]$. Integrating (2.1) over $[t_i, t_{i+1}]$ yields

$$y(t_{i+1}) = y(t_i) + \int_{t_i}^{t_{i+1}} f(y, t)dt , \tag{2.3}$$

which we approximate by

$$y_{i+1} = y_i + \int_{t_i}^{t_{i+1}} f(y_{l_1, \ldots, l_d}, t)dt , \tag{2.4}$$

where $y_{l_1, \ldots, l_d}$ is some generalized Hermite interpolant of $y$ defined by

$$y_{l_1,\ldots, l_d} \in S^h \equiv \left\{ t^n \exp(\lambda_m t) \mid n = 0, \ldots, N_m - 1 ; \quad m = 1, \ldots, M ; \sum_m N_m = l \right\} , \tag{2.5}$$

subject to the following interpolation conditions at $t_{ij} = t_i + \tau_j h_i$

$$D^{k-1} y_{l_1,\ldots, l_d}(t_{ij}) = D^{k-1} y(t_{ij}) , \quad j = 1, \ldots, d ; \quad k = 1, \ldots, l_j , \tag{2.6}$$

with $0 \leqslant \tau_1 < \ldots < \tau_d \leqslant 1$. Using collocation (eventually multiple), the derivatives are eliminated by

$$D^{k-1} y(t_{ij}) = D^{k-2} f(y(t_{ij}), t_{ij}), \ k = 2, \ldots, l_j . \tag{2.7}$$

As a result, $y_{l_1,\ldots,l_d}$ in (2.4) only depends on $y_i$ if $\tau_1 = 0$, $y_{ij}$ for all the $\tau_j'$ s comprised between 0 and 1, and $y_{i+1}$ if $\tau_d = 1$. $y_i$ is known from the initial condition (2.2) if $i = 0$ or from previous time integration if not. The intermediate $y_{ij}$'s are given by

$$y_{ij} = y_i + \int_{t_i}^{t_{ij}} f(y_{l_1,\ldots,l_d}, t)dt \quad , \tag{2.8}$$

while $y_{i+1}$ is given by (2.4). In (2.4) as in (2.8), the integral terms are normally approximated by some quadrature formula, eventually a different one for each $j$. The final result is a system of nonlinear equations to be solved for $y_{ij}$ and $y_{i+1}$, or equivalently a single nonlinear equation in $y_{i+1}$ reading

$$y_{i+1} = y_i + h_i \emptyset_i (y_i, y_{i+1}) \quad . \tag{2.9}$$

$y_{i+1}$ exists and is unique if the second member of (2.9) can be shown to be a contraction.

In [7], a complete analysis of the above class of schemes has been given with "ad hoc" demonstrations in the special case where $d = 2$ with $\tau_1 = 0$ and $\tau_2 = 1$. We want to show here how the Tchebycheffian systems theory given in [8] can be used to provide more general results than in [7] and therefore leads to a more comprehensive understanding of the schemes proposed hereabove. In parallel to theorem 1 of Ref. 7, we have

*THEOREM 1*

*There is one and only one generalized Hermite interpolant* $y_{l_1,\ldots,l_d} \in S^h$ *verifying the interpolation conditions* (2.6)

*PROOF*

First, we must show that $S^h$ is an *Extended Complete Tchebycheff* (ECT)-space. To show this, we need only display a canonical ECT-system that forms a basis for it. Such a canonical ECT-system can be defined as in [8, p. 364] with respect to the weight functions

$$\exp(\lambda_1 t), \underbrace{1, 1, \ldots, 1}_{(N_1 - 1)\text{times}}, \qquad \exp(\lambda_2 - \lambda_1)t, \underbrace{1, 1, \ldots, 1,}_{(N_2 - 1)\text{times}}$$

$$\ldots, \exp(\lambda_M - \lambda_{M-1})t, \underbrace{1, 1, \ldots, 1}_{(N_{M-1})\text{times}}$$

and if we call $u_i$, $i = 1, \ldots, l$, the resulting basis functions, it is easy to see that $S^h = \text{span}\,\{u_i\}$ so that we have shown that $S^h$ is an ECT-space. Theorem 1 is then a direct reformulation of the first part of Theorem 9.9 in [8].

As in the Appendix of [7], it can be proved that $y_{l_1,\ldots,l_d}$ satisfies a Lipschitz condition with respect to $y_{i+1}$, with a Lipschitz constant independent of h. The existence of a unique solution to (2.10) is then guaranteed for h sufficiently small by an argument identical to the one used in [2, Section 2].

Discrete error bounds can now be derived for the methods presented hereabove as soon as the approximation order of the interpolation in $S^h$ is known. This result is contained in

*LEMMA 1*

*Let* $y \in C^k[0,T]$ *with* $k \geq l$ *and let* $\tilde{y}_{l_1,\ldots,l_d} \in S^h$ *be its unique generalized Hermite interpolant over* $[t_i, t_{i+1}]$ *verifying the interpolation conditions* (2.6). *Then*

$$|y(t) - \tilde{y}_{l_1,\ldots,l_d}(t)| \leq C_1 h_i^l \,, \quad \forall\, t \in (t_i, t_{i+1}) \tag{2.10}$$

This Lemma is a direct consequence of the second part of theorem 9.9 and of Lemma 9.6 in [8]. With its help, it is possible to prove

*THEOREM 2*

*Assume that* $f(y,t)$ *is of class* $C^k$, $k \geq l-1$, *over* $R \times [0,T]$, *and let the discrete one-step integration method be defined as above for* $y_{l_1,\ldots,l_d} \in S^h$. *Then there exists a constant* $C$ *such that*

$$|y(t_i) - y_i| \leq C h^l \,, \quad i = 1, \ldots, N \,, \tag{2.11}$$

*under the hypotheses that the numerical quadratures used over* $[t_i, t_{i+1}]$ *(resp.* $[t_i, t_{ij}]$*) are of order* $(l + 1)$ *(resp.* $l$*).*

*PROOF*

The local truncation error is given by

$$\varepsilon_i = y(t_{i+1}) - y_{i+1}$$

$$= \int_{t_i}^{t_{i+1}} f(y, t)dt - \int_{t_i}^{t_{i+1}} f(\tilde{y}_{l_1, \ldots, l_d}, t)dt$$

$$+ \int_{t_i}^{t_{i+1}} f(\tilde{y}_{l_1, \ldots, l_d}, t)dt - \int_{t_i}^{t_{i+1}} f(y_{l_1, \ldots, l_d}, t)dt$$

$$+ \int_{t_i}^{t_{i+1}} f(y_{l_1, \ldots, l_d}, t)dt - h_i \sum_{k=1}^{n} w_k f(y_{l_1, \ldots, l_d}(t_{ik}), t_{ik})$$

$$= \varepsilon_i^{(1)} + \varepsilon_i^{(2)} + \varepsilon_i^{(3)}, \qquad \textbf{(2.12)}$$

where $\tilde{y}_{l_1, \ldots, l_d}$ is the generalized Hermite interpolant of y introduced in Lemma 1, and the $t_{ik}$'s are quadrature points over $[t_i, t_{i+1}]$. Using the Lipschitz property of f and Lemma 1,

$$|\varepsilon| \leq C_1 h_i^{l+1} + L \int_{t_i}^{t_{i+1}} |\tilde{y}_{l_1, \ldots, l_d} - y_{l_1, \ldots, l_d}| dt$$

$$+ e_{i+1} \qquad \textbf{(2.13)}$$

where $e_{i+1} = |\varepsilon_i^{(3)}|$ is the (absolute) quadrature error over $[t_i, t_{i+1}]$. Similarly for each intermediate $t_{ij}$,

$$|y(t_{ij}) - y_{ij}| \leq C_1 h_i^{l+1} + L \int_{t_i}^{t_{i+1}} |\tilde{y}_{l_1, \ldots, l_d} - y_{l_1, \ldots, l_d}| dt$$

$$+ e_j, \qquad \textbf{(2.14)}$$

where $e_j$ is the (absolute) quadrature error over $[t_i, t_{ij}]$. From (2.14), one obtains

$$\sup_{j=1,\ldots,d} |y(t_{ij}) - y_{ij}| \leq C_1 h_i^{l+1} + h_i L \sup_{t \in [t_i, t_{i+1}]} |\tilde{y}_{l_1, \ldots, l_d}(t) - y_{l_1, \ldots, l_d}(t)|$$

$$+ \sup_{j=1,\ldots,d} e_j . \qquad \textbf{(2.15)}$$

Since $\tilde{y}_{l_1,\ldots,l_d} - y_{l_1,\ldots,l_d}$ is an element of $S^h$ which depends linearly on the arguments $D^{k-1}(y(t_{ij}) - y_{ij})$, $j = 1, \ldots, d$, $k = 1, \ldots, l_j$ where

$$\left|D^{k-1}(y(t_{ij}) - y_{ij})\right| = \left|D^{k-2}\left\{f(y(t_{ij}), t_{ij}) - f(y_{ij}, t_{ij})\right\}\right|$$
$$\leqslant L_{k-2}\left|y(t_{ij}) - y_{ij}\right| , \tag{2.16}$$

there exists a constant $C_2$ such that

$$\sup_{t\in[t_i,\, t_{i+1}]} \left|\tilde{y}_{l_1,\ldots,l_d}(t) - y_{l_1,\ldots,l_d}(t)\right| \leqslant C_2 \sup_{j=1,\ldots,d} \left|y(t_{ij}) - y_{ij}\right| \tag{2.17}$$

Consequently

$$|\varepsilon_i^{(2)}| \leqslant C_2 h_i L \sup_{j=1,\ldots,d} |y(t_{ij}) - y_{ij}| ,$$

which becomes by combining (2.15) and (2.17)

$$|\varepsilon_i^{(2)}| \leqslant \frac{C_2 h_i L}{1-C_2 h_i L}\left(C_1 h_i^{l+1} + \sup_{j=1,\ldots,d} e_j\right) \tag{2.18}$$

As a final result, there exist constants $C_\alpha$ and $C_\beta$ such that for $h$ sufficiently small

$$|\varepsilon_i| \leqslant C_\alpha h^{l+1} + C_\beta h \sup_{j=1,\ldots,d} e_j + e_{i+1} \tag{2.19}$$

If the error of quadrature is assumed to be of $O(h^{l+1})$ (resp. $O(h^l)$) over the entire interval $[t_i, t_{i+1}]$ (resp. the partial interval $[t_i, t_{ij}]$), $|\varepsilon_i|$ is of $O(h^{l+1})$ and the bound in (2.11) is a direct consequence of Henrici's theorem 2.2[9].

From the discrete error bounds given in Theorem 2, continuous error bounds could also be derived as in Ref. 2, but we shall not pursue this point here.

## 3 - SOME EXAMPLES

For stiff differential equations arising for instance from the semi discretization of parabolic PDE'S, the common choices for $S^h$ consist of pure polinomial basis functions ($S^h \equiv P_{l-1}$, the space of polynomials of degree not greater than $l-1$) or of a mixture of polynomials and exponential basis functions with real negative $\lambda_m'$s . The special cases where $d = 2$ with $\tau_1 = 0$ and $\tau_2 = 1$ have been treated in [2] for $P_{l-1}$ and in [7] for the general space $S^h$.

In the presence of pure polynomial basis functions but with intermediate points ($0 < \tau_j < 1$), the formalism proposed here can be seen as a new presentation of the block-implicit methods [10-12] viewed as one step methods. Extensions to more general spaces $S^h$ naturally provide exponential fitting capabilities which were not present in the original block implicit formalism. Assume for instance that $y$, the exact solution, belongs to $S^h$ : then it is identical to its interpolation in $S^h$ in virtue of Theorem 1 ($y \equiv y_{l_1,\ldots,l_d}$) and since $f(y, t) = Dy$ also belongs to $S^h$ (since $DS^h \leqslant S^h$), numerical quadrature rules which are exact for members of $S^h$ (and consequently of $O(h^{l+1})$) will not introduce any error in (2.4) and (2.8). In other words, if $y \in S^h$, it will be integrated exactly by these methods provided the quadrature rules are exact for members of $S^h$. In view of Theorem 2, less accuracy is required for the intermediate integrals (2.8) and we can correspondingly relax somewhat the above prescriptions in the sense that the intermediate quadrature rules may only be exact for members of $DS^h$, a clear advantage when $\dim(DS^h) < \dim(S^h)$ in particular when one at least of the $\lambda_m$'s is equal to zero. The schemes proposed hereabove are thus exponentially fitted of order $N_m-1$ at $\lambda_m$, $m = 1,\ldots, M$ [13].

## REFERENCES


[1] G. FAIRWEATHER, Finite Element Galerkin Methods for Differential Equations, Marcel Dekker, New York (1978)

[2] J.P. HENNART, "One-step piecewise polynomial multiple collocation methods for initial value problems", Mathematics of Computation, 31, pp. 24-36 (1977)

[3] B. BJUREL, G. DAHLQUIST, B. LINDBERG, S. LINDE and L. ODEN, "Survey of stiff ordinary differential equations", Report NA 70.11, The Royal Institute of Technology, Stockholm, Sweden (1972)

[4] J.P. HENNART, "Multiple collocation finite elements in time for parabolic evolution problems", in The Mathematics of Finite Elements and Applications III MAFELAP 1978, pp. 271-278, J.R. Whiteman, Ed., Academic press, London (1979)

[5] J.P. HENNART, "On the implementation of finite elements for parabolic evolution problems", to appear in The Mathematics of Finite Elements and Applications IV, MAFELAP 1981, J.R. Whiteman, Ed., Academic Press, London. Also available as Comunicaciones Técnicas, Série NA, No 264, IIMAS-UNAM (1981)

[6] J.P. HENNART, "Topics in finite element discretization of parabolic evolution problems". In these Proceedings.

[7] J.P. HENNART and H. GOURGEON, "One-step exponentially fitted piecewise continuous methods for initial value problems", Comunicaciones Técnicas, Série NA, No 263, IIMAS-UNAM (1981)

[8] L. SCHUMAKER, Spline Functions : Basic Theory, John Wiley & Sons, New York, (1981)

[9] P. HENRICI, Discrete Variable Methods in Ordinary Differential Equations, J. Wiley & Sons, New York (1962).

[10] L.F. SHAMPINE and H.A. WATTS, "Block implicit one-step methods", Mathematics of Computation, 23, pp. 731-740 (1969)

[11] H.A. WATTS and L.F. SHAMPINE, "A stable block implicit one-step methods", BIT, 12, pp. 252-266 (1972)

[12] D.S. WATANABE, "Block implicit one-step methods", Mathematics of Computation, 32, pp. 405-414 (1978)

[13] G. BJUREL, "Modified linear multistep methods for a class of stiff ordinary differential equations", BIT, 12, pp. 142-160 (1972).

# Recursive Quadratic Programming Algorithms and their Convergence Properties

R.W.H. Sargent

## Synopsis

Local superlinear convergence of the recursive quadratic programming algorithm is proved under weaker conditions than those used by previous workers.
It is also shown that a modified Lagrangian function, which contains no arbitrary parameters, can be used as a descent function to force global convergence under reasonable conditions, and that use of this function does not impair the final superlinear convergence if this would otherwise be achieved.
Variants of the algorithm are also given, which first concentrate on finding a feasible point and thereafter maintain feasibility.

Presented at the Third IIMAS Workshop on Numerical Analysis, Cocoyoc, Mexico, 11-16 January, 1981.

Department of Chemical Engineering,
Imperial College,
London SW7 2BY.

# Recursive Quadratic Programming Algorithms and their Convergence Properties

R.W.H. Sargent
(Imperial College, London)

## 1. Introduction

Among the many approaches to solving the general nonlinear programming problem (see for example, 1), the most promising seems to be that based on "recursive quadratic programming". In this method a sequence of points is generated, each of which is a solution of a quadratic programme, formed by making a linear approximation to the constraints and a quadratic approximation to the Lagrangian function for the problem at each successive point. It seems to have been proposed first by Wilson (2) in an unpublished thesis, but current interest stems from the work of Han (3,4), who gave some local convergence results and also proposed a method of forcing global convergence under reasonable conditions by using the solution of the quadratic programme at each step to define a search direction, and minimizing an "exact penalty function" along this direction.

Han's original algorithm gave rise to a number of difficulties, but other workers, notably Mayne and Polak (5) and Chamberlain et al (6), have proposed practical variants which circumvent these difficulties. Nevertheless, the resulting algorithms are relatively complicated, and there is some doubt about the efficacy of the technique for choosing the weighting parameters in the exact penalty function; there is some numerical evidence to show that unduly high values cause slow convergence, yet there is so far no known mechanism for reducing the weighting parameters as the iterations proceed without destroying the global convergence property.

In the present paper we show that a modified Lagrangian function, which contains no arbitrary parameters, can be used as a descent function to force global convergence under reasonable conditions.

Using the approach of Dennis and Walker (7,8) for unconstrained minimization, we prove local superlinear convergence of the recursive quadratic programming algorithm, when the Hessian matrix of the objective function is either that of the Lagrangian function or a quasi-Newton approximation to it. The conditions used are significantly weaker than those of earlier workers (3,5,6,9,10), and we also show that use of the modified Lagrangian as a descent function does not impair the superlinear rate of convergence.

Finally, we give variants of the algorithm which first concentrate on obtaining a feasible point, and thereafter reduce the descent function while maintaining feasibility. These variants are a natural development from earlier recursive quadratic programming algorithms (11).

## 2. The QP Subproblem

For a general computer program the most convenient form of the nonlinear programming problem is:

$$\left.\begin{array}{llll} \text{Minimize} & f^{0}(x) & , & \text{(i)} \\ \text{subject to} & a \leqslant x \leqslant b & , & \text{(ii)} \\ \text{and} & c \leqslant f(x) \leqslant d & , & \text{(iii)} \end{array}\right) \quad (2.1)$$

where $x \varepsilon R^n$, $f^0 : R^n \to R$, $f : R^n \to R^m$. In this formulation unbounded elements can be accommodated by setting the relevant bound to machine limits, while equality constraints can be imposed by setting the relevant upper and lower bounds to the same value.

It is easy to ensure that all points considered satisfy the linear constraints in (2.1), which define a convex subset $X \subset R^n$, so in practice the function $f^0(x)$ and the nonlinear elements of $f(x)$ need only be defined on X. We shall assume that X is compact with non-empty interior, and that $f^0(x)$ and $f(x)$ are continuously differentiable on X.

The constraint set is said to be strictly regular at $x \varepsilon X$ if the only vectors $\lambda \varepsilon R^m$, $\mu \varepsilon R^n$ which satisfy the conditions:

$$\left.\begin{array}{llll} & \mu + \lambda . f_x(x) = 0 & & \\ \mu^i(a^i - x^i) \geqslant 0 , & \mu^i(b^i - x^i) \geqslant 0 & , & i=1,2, \ldots n, \\ \lambda^j(c^j - f^j(x)) \geqslant 0 , & \lambda^j(d^j - f^j(x)) \geqslant 0 & , & j=1,2, \ldots m, \end{array}\right) \quad (2.2)$$

are $\lambda = 0$, $\mu = 0$. We note that this definition is not restricted to feasible points, but if x is feasible it corresponds to the Mangasarian-Fromovitz constraint qualification (see reference 12).

If the constraints are strictly regular at the solution of (2.1), then this solution satisfies the Kuhn-Tucker conditions:

$$\left.\begin{array}{llll} & f^o_x(x) = \mu + \lambda . f_x(x) & , & \text{(i)} \\ \mu^i(a^i - x^i) \geqslant 0 , & \mu^i(b^i - x^i) \geqslant 0 & , \ i=1,2, \ldots n, & \text{(ii)} \\ \lambda^j(c^j - f^j(x)) \geqslant 0 , & \lambda^j(d^j - f^j(x)) \geqslant 0 & , \ j=1,2, \ldots m, & \text{(iii)} \end{array}\right) \quad (2.3)$$

for some $\lambda \varepsilon R^m$, $\mu \varepsilon R^n$, as well as the constraints (2.1.ii and iii). In practice we we do not attempt to solve (2.1), but assume strict regularity and seek a Kuhn-Tucker point. In fact we shall also be content to satisfy the nonlinear constraints {NL} in (2.1) to within a prescribed tolerance $\varepsilon$, yielding the "relaxed" problem:

$$\left.\begin{array}{llll} \text{Minimize} & f^0(x) & , & \text{(i)} \\ \text{subject to} & a \leqslant x \leqslant b & , & \text{(ii)} \\ & c^j \leqslant f^j(x) \leqslant d^j & , \ j \notin NL , & \text{(iii)} \\ & c^j - \varepsilon \leqslant f^j(x) \leqslant d^j + \varepsilon & , \ j \varepsilon NL , & \text{(iv)} \end{array}\right) \quad (2.1a)$$

with Kuhn-Tucker conditions:

$$\left.\begin{array}{llll} f^o_x(x) = \mu+\lambda.f_x(x) & & & (i) \\ \mu^i(a^i-x^i)\geq 0 & , \mu^i(b^i-x^i)\geq 0 & , i=1,2, \ldots n, & (ii) \\ \lambda^j(c^j-f^j(x))\geq 0 & , \lambda^j(d^j-f^j(x))\geq 0 & , j\notin NL & ,(iii) \\ \lambda^j(c^j-\varepsilon-f^j(x))\geq 0 & , \lambda^j(d^j+\varepsilon-f^j(x))\geq 0 & , j\varepsilon NL & , (iv) \end{array}\right) \quad (2.3a)$$

We shall write $z\equiv(x,\lambda,\mu)$ and use $Z^*$ to denote the set of z representing solutions of (2.3a). The set $Z^*$ is non-empty if there is a solution of (2.1a) at which the constraints are strictly regular.

If the functions $f^0(x)$, $f(x)$ are in fact twice continuously differentiable, Taylor expansions of these functions yield the following linear approximation to (2.3):

$$\left.\begin{array}{llll} & f^o_x(x) + p.H = \mu + \lambda.f_x(x) & & \\ a^i\leq x^i+p^i\leq b^i & , \mu^i(a^i-x^i-p^i)\geq 0 & , \mu^i(b^i-x^i-p^i)\geq 0 & , i=1,2, \ldots n \\ c^j\leq\phi^j(x,p)\leq d^j & , \lambda^j(c^j-\phi^j(x,p))\geq 0 & , \lambda^j(d^j-\phi^j(x,p))\geq 0 & , j=1,2, \ldots m \end{array}\right) \quad (2.4)$$

where $\phi(x,p) = f(x) + f_x(x).p$

and $$H = f^o_{xx}(x) - \lambda.f_{xx}(x). \quad (2.4a)$$

However, it is not necessary for the success of the algorithms to be described that H satisfy (2.4a), and in general we shall treat H as independently defined.

Conditions (2.4) are the Kuhn-Tucker conditions for the quadratic programme (QP):

$$\left.\begin{array}{lll} \text{Minimize} & f_x(x).p + \tfrac{1}{2}p.H.p & , \\ \text{subject to} & a\leq x+p\leq b & , \\ \text{and} & c\leq\phi(x,p)\leq d & , \end{array}\right) \quad (2.5)$$

We shall denote this problem by $Q_0(x,H)$, with solution set $Z_0(x,H)$, where $z = \{x+p,\lambda,\mu\}$. If the constraint set for problem (2.1) is strictly regular at $x\varepsilon X$, then the constraint set for $Q_0(x,H)$ is strictly regular at every feasible point for $Q_0(x,H)$, and we then say that $Q_0(x,H)$ is strictly regular (see reference 12, Theorem 7).

It will be important for the algorithms that the QP subproblems be strictly regular, and although this cannot be achieved in all circumstances, the chances are improved by use of a modification, similar to that used by Powell (13), as follows:

For each $j\varepsilon NL$ and some specified $\theta\varepsilon(0,1)$:

$$\left.\begin{array}{lll} \text{(i)} \quad \text{Set } \bar{c}^j_o = c^j-f^j(x) & , \bar{d}^j_o = d^j-f^j(x) & , \\ \qquad \Delta c^j_o = \max(\varepsilon,\bar{c}^j_o) & , \Delta d^j_o = \max(\varepsilon,-\bar{d}^j_o) & . \\ \text{(ii)} \quad \text{For } r=1,2, \ldots. \text{ compute} & & \end{array}\right) \quad (2.6)$$

$$\Delta c_r^j = \theta.\Delta c_{r-1}^j \quad , \quad \bar{c}_r^j = \bar{c}_{r-1}^j - (1-\theta).\Delta c_{r-1}^j \, , \qquad (2.6)$$
$$\Delta d_r^j = \theta.\Delta d_{r-1}^j \quad , \quad \bar{d}_r^j = \bar{d}_{r-1}^j + (1-\theta).\Delta d_{r-1}^j \, .$$

This procedure yields

$$\bar{c}_r - \Delta c_r = \bar{c}_o - \Delta c_o \leqslant 0 \, , \qquad (2.7)$$
$$\bar{d}_r - \Delta d_r = \bar{d}_o - \Delta d_o \geqslant 0 \, ,$$

with $\lim_{r\to\infty} \Delta c_r = 0$ , $\lim_{r\to\infty} \Delta d_r = 0$.

We also write $\bar{a} = a-x$ , $\bar{b} = b-x$

$$\bar{c}_r^j = c^j - f^j(x) \, , \quad \bar{d}_r^j = d^j - f^j(x), \; j \notin NL, \text{ all } r, \qquad (2.6a)$$

yielding the modified subproblem $Q_r(x,H)$:

$$\text{Minimize} \quad f_x(x).p + \tfrac{1}{2}p.H.p \, , \qquad (2.8)$$
$$\text{subject to} \quad \bar{a} \leqslant p \leqslant \bar{b} \, ,$$
$$\text{and} \quad \bar{c}_r \leqslant f_x(x).p \leqslant \bar{d}_r \, ,$$

with solution set $Z_r(x,H)$.

It is easy to see that if p=0 is a solution to (2.8) and H is bounded, then any $z \in Z_r(x,H)$ is a solution of (2.3a), or $Z_r(x,H) \subset Z^*$.

For the linear equality constraints in (2.1) it is always possible to reject redundant constraints, leaving a linearly independent set defining the same feasible region. For our analysis we shall assume that this has been done, although if a Kuhn-Tucker point exists many practical QP algorithms will handle redundant equality constraints without difficulty. Strict regularity of $Q_r(x,H)$ is then ensured if its constraint set has a non-empty relative interior.

The advantages and deficiencies of procedure (2.6) in securing regularity are illustrated by the following example:

Example

$$f_1(x) = (x_1-1)^2 + (x_2-1)^2 \geqslant 4 \, , \qquad (a)$$
$$f_2(x) = x_1+x_2 \leqslant 5 \, , \qquad (b)$$
$$x_1 \geqslant 0 \, , \; x_2 \geqslant 0 \, . \qquad (c)$$

Linearizing about the point (2,2) yields

$$x_1+x_2 \geqslant 6 \qquad (a)$$
$$x_1+x_2 \leqslant 5 \qquad (b)$$
$$x_1 \geqslant 0 \, , \; x_2 \geqslant 0 \qquad (c)$$

which are of course inconsistent. However, as r increases the constant on the right-hand side of (a) decreases from 6 to 4 in the limit, thus yielding a region with non-empty interior for suitably large r.

On the other hand, linearizing about the point $(-\frac{1}{3}, -\frac{1}{3})$ yields

$$x_1 + x_2 \leqslant -5/6 \quad , \qquad (a)$$

$$x_1 + x_2 \leqslant 5 \quad , \qquad (b)$$

$$x_1 \geqslant 0,\ x_2 \geqslant 0 \quad , \qquad (c)$$

which are again inconsistent. As r increases, the constant on the right-hand side of (a) increases from -5/6 to -2/3, so in this case the constraints remain inconsistent and the procedure fails.

## 3. The Basic Algorithm

The essence of the proposed algorithm is to solve a sequence of QP subproblems $Q_r(x_k,H_k)$ , k=0,1,2, .... , using the solution $p_k$ of $Q_r(x_k,H_k)$ to generate the next point in the sequence according to the relation

$$x_{k+1} = x_k + \alpha_k p_k \ , \qquad (3.1)$$

where $\alpha_k$ is a scalar.

This defines the general class of recursive quadratic programming algorithms, and to obtain a specific algorithm we need rules for the choice of $\alpha_k$ and $H_k$ at each step.

We first observe that, since X is convex and the linear constraints defining X are included in the constraint set for each $Q_r(x_k,H_k)$, we can ensure that the whole sequence $\{x_k\}$, k=0,1,2, .... remains in X by choosing $x_0 \varepsilon X$ and $\alpha_k \varepsilon (0,1]$ for each k.

At each step we use the procedure defined in (2.6) to choose the smallest r for which $Q_r(x_k,H_k)$ is regular. If a finite r does not exist, the algorithm fails, but otherwise this procedure, together with the choice $\alpha_k \varepsilon (0,1]$, ensures that the sequence $\{z_k\}$ is well defined provided that the $H_k$ remain bounded.

The reason for introducing the scalar $\alpha_k$ is to ensure convergence of the sequence $\{x_k\}$ to a Kuhn-Tucker point of problem (2.1a). Now in the unconstrained case (see for example reference 14) convergence can be forced by finding a descent direction $p_k$ for the objective function $f^0(x)$ at point $x_k$, then choosing $\alpha_k$ in (3.1) to satisfy the "descent condition":

$$f^0(x_k) - f^0(x_{k+1}) \geqslant -\delta . \alpha_k f^0_x(x_k).p_k \qquad (3.2)$$

for some fixed $\delta \varepsilon (0,1)$. The descent direction $p_k$ is often generated using some positive-definite matrix $H_k$ in the equation

$$p_k.H_k = -f_x(x_k), \qquad (3.3)$$

and $H_k$ is often chosen to approximate the second derivative of the objective function $f^0_{xx}(x)$.

In the constrained case, we see from (2.4a) that it is desirable for the $H_k$ in our sequence to approximate the second derivatives of the Lagrangian function. It is therefore natural to think of using this function as the descent function in the constrained case, as was indeed proposed by Powell et al (13,6). However it turns out that the solution $p_k$ of the QP subproblem is not necessarily a descent direction for the Lagrangian function itself, and instead we shall show that we can use the modified Lagrangian function:

$$\left.\begin{aligned} & L(r,\alpha,z) = f^o(x) + \alpha(\mu.\Delta x_r + \lambda.\Delta f_r) \quad , \\ \text{where} \quad \text{if} \quad & \mu^i \geqslant 0 : \Delta x_r^i = a^i - x^i = \bar{a}^i \quad , \\ & \mu^i < 0 : \Delta x_r^i = b^i - x^i = \bar{b}^i \quad , \\ & \lambda^j \geqslant 0 : \Delta f_r^j = \bar{c}_r^j \quad , \\ & \lambda^j < 0 : \Delta f_r^j = \bar{d}_r^j \quad . \end{aligned}\right\} \quad (3.4)$$

If $z \varepsilon Z_r(x,H)$, it follows from (3.4) and the Kuhn-Tucker conditions for $Q_r(x,H)$ that

$$\begin{aligned} L(r,\alpha,z) &= f^o(x) + \alpha(\mu.p + \lambda.f_x(x).p) \\ &= f^o(x) + \alpha(f_x^o(x).p + p.H.p). \end{aligned} \quad (3.5)$$

We also have, using (2.6)-(2.8) in (3.4):

$$\begin{aligned} L(r_{k+1},\alpha_{k+1},z_{k+1}) &\leqslant f^o(x_{k+1}) + \alpha_{k+1}(\mu_{k+1}.\Delta x_{o,k+1} + \lambda_{k+1}.\Delta f_{o,k+1}) \\ &\leqslant f^o(x_{k+1}) + \alpha_{k+1}(\mu_{k+1}-\mu_k).\Delta x_{o,k+1} + (\lambda_{k+1}-\lambda_k).\Delta f_{o,k+1}) \\ &\quad + \alpha_{k+1}(\mu_k.\Delta x_{o,k+1} + \lambda_k \Delta f_{o,k+1}) \\ &\leqslant f^o(x_{k+1}) + \alpha_{k+1}.\Delta_{k+1} + o(\| \alpha_k p_k \|) \end{aligned} \quad (3.6)$$

where $\Delta_{k+1} = (\mu_{k+1}-\mu_k).\Delta x_{o,k+1} + (\lambda_{k+1}-\lambda_k)\Delta f_{o,k+1} + (1-\alpha_k)(\mu_k \Delta x_{o,k} + \lambda_k.\Delta f_{o,k}).$ (3.7)

Hence, writing $L_k \equiv L(r_k,\alpha_k,z_k)$ etc., (3.5) and (3.6) yield

$$L_k - L_{k+1} \geqslant \alpha_k p_k.H_k.p_k - \alpha_{k+1}\Delta_{k+1} + o(\| \alpha_k p_k \|). \quad (3.8)$$

We now choose an upper bound $\bar{\alpha}_{k+1}$ for $\alpha_{k+1}$ as follows:

$$\left.\begin{aligned} &\text{if } \Delta_{k+1} \leqslant 0 \text{ , set } \bar{\alpha}_{k+1} = 1 \\ &\text{if } \Delta_{k+1} > 0 \text{ , set } \bar{\alpha}_{k+1} = \min(1, \delta'.\alpha_k p_k H_k p_k / \Delta_{k+1}) \text{ ,} \end{aligned}\right\} \quad (3.9)$$

for some specified $\delta' \varepsilon (0,1)$. We also set $\bar{\alpha}_o = 1$.

Then for any $\alpha_{k+1} \varepsilon (0,\bar{\alpha}_{k+1}]$, and any $\delta \varepsilon (0,1-\delta')$, it follows that there is an $\alpha_k' \varepsilon (0,\bar{\alpha}_k]$ such that for all $\alpha_k \varepsilon (0,\alpha_k']$ we have $L_k - L_{k+1} \geqslant \delta.\alpha_k p_k.H_k.p_k$. (3.10)

To choose a suitable $\alpha_k$ we can use the Armijo rule, $\alpha_k = \theta^s \bar{\alpha}_k$, where $\theta \varepsilon (0,1)$ and s is the smallest non-negative integer for which (3.10) is satisfied.

We then have the following global convergence theorem:

Theorem 1

Suppose that for each $x \varepsilon X$ the subproblem $Q_r(x,H)$ is regular for some finite r and that the sequence $\{x_k\}$ is generated as above using a bounded sequence $\{H_k\}$ such that $p_k.H_k.p_k \geq \rho \| p_k \|^2$ for all k and some fixed scalar $\rho > 0$. Then the set Z* is non-empty, and the sequence $\{z_k\}$ either attains a point in Z* or converges to it in the sense that $\lim_{k\to\infty} \inf_{z^* \varepsilon Z^*} \| z^* - z_k \| = 0$. Moreover if the Kuhn-Tucker points of (2.1a) are isolated, the algorithm converges to one of these points.

Proof

First suppose that Z* is empty. Since $f^0(x)$ and $f(x)$ are continuous on X, the set of feasible points for problem (2.1a) is closed, and if it is non-empty $f^0(x)$ attains its minimum on the set. But since Z* is empty, the constraints cannot then be regular at the minimizing point, and this implies that the subproblem $Q_r(x,H)$ based on this point cannot be regular for any finite r. On the other hand, if the feasible set for problem (2.1a) is empty, then again there must be at least one $x \varepsilon X$ for which $Q_r(x,H)$ cannot be regular for any finite r. Thus Z* must in fact be non-empty.

We have already seen that under the hypotheses of the theorem the sequence $\{x_k\}$ remains in X, and the sequence $\{z_k\}$ is well defined. If $\{z_k\}$ attains a point in Z* there is nothing more to prove, so we proceed to consider the case when $z_k \notin Z^*$, all k.

Suppose that $\lim_{k\to\infty} p_k \neq 0$. Then there is an infinite subsequence, $\{p_i\}$ say, with limit point $\bar{p} \neq 0$, and an index $i_o$ such that $\| p_i \| \geq \frac{1}{2} \| \bar{p} \| > 0$ for all $i \geq i_o$, so that

$$p_i.H_i.p_i \geq \rho \| p_i \|^2 \geq \tfrac{1}{4}\rho \| \bar{p} \|^2 > 0, \qquad i \geq i_o \tag{3.11}$$

Now from the mean value theorem we have

$$f^0(x_{i+1}) - f^0(x_i) = \alpha_i f^0_x(x_i + \beta\alpha_i p_i).p_i$$

for some $\beta \varepsilon [0,1]$. But since X is compact, the sequence $\{p_k\}$ is uniformly bounded and $f^0_x(x)$ is uniformly continuous on X, so for each $\varepsilon > 0$ there is a fixed $\hat{\alpha} \varepsilon (0,1]$ such that for any $\alpha_i \varepsilon (0,\hat{\alpha}]$ and any $\beta \varepsilon [0,1]$:

$$|f^0_x(x_i + \beta\alpha_i p_i).p_i - f^0_x(x_i).p_i| \leq \varepsilon,$$

and hence $\quad |f^0(x_{i+1}) - f^0(x_i) - \alpha_i f^0_x(x_i) p_i| \leq \varepsilon.$

The same argument applies to each element $f^j(x)$, j=1,2, ... m, and $\varepsilon$ can be chosen so that $(m+1)\varepsilon \leq \frac{1}{4}\rho(1-\delta'-\delta) \| \bar{p} \|^2$. It then follows that the terms of $o(\| \alpha_k p_k \|)$ in (3.8) are such that (3.10) is satisfied for all $i \geq i_o$ and any $\alpha_i \varepsilon (0,\hat{\alpha}]$, and the use of the Armijo rule to determine $\alpha_i$ ensures that $\alpha_i \geq \theta\hat{\alpha}$. Hence

$$L_i - L_{i+1} \geq \delta.\theta\hat{\alpha}.\tfrac{1}{4}\rho \| \bar{p} \|^2 > 0, \quad i \geq i_o. \tag{3.12}$$

But this implies that $\{L_i\}$, $i \geq i_o$, is a monotonically decreasing sequence, and from

(3.5) this sequence has a uniform lower bound, so that $(L_i - L_{i+1}) \to 0$. Since this contradicts (3.12), we must in fact have $p_k \to 0$.

Now suppose that $\{z_k\}$ does not tend to $Z^*$. Then there exists an infinite subsequence, $\{z_i\}$ say, with limit point $\bar{z} \notin Z^*$, and from above $\lim_{i\to\infty} p_i = \bar{p} = 0$. But, as noted in Section 2, this implies that $\bar{z} \in Z^*$, again producing a contradiction. Thus $\{z_k\}$ must tend to $Z^*$.

The proof that $p_k \to 0$ implies convergence to one of the Kuhn-Tucker points if these are isolated is as given by Ortega and Rheinboldt {15, p476, §14.1.5.}.

Q.E.D.

The proof is a straightforward extension of a proof given previously for the unconstrained case by Polak, Sebastian and Sargent {14}. It follows from the theorem that the algorithm will terminate in a finite number of steps if it is terminated when $\| H_k p_k \| \leqslant \varepsilon$, for any $\varepsilon > 0$.

## 4. Choice of $H_k$, and the Rate of Convergence

Following Wilson (2), we could compute $H_{k+1}$ as the Hessian matrix of the Lagrangian (2.4a), using the latest approximation to the solution:

$$H_{k+1} = f^o_{xx}(x_{k+1}) - \lambda_k f_{xx}(x_{k+1}). \tag{4.1}$$

However it will usually be more convenient to use a quasi-Newton approximation, based on gradient changes with respect to x:

$$q_k^T = (f^o_x(x_{k+1}) - f^o_x(x_k)) - \lambda_k(f_x(x_{k+1}) - f_x(x_k)). \tag{4.2}$$

In the practical implementation of the algorithm we use the well known BFGS formula, in the form:

$$H_{k+1} = H_k + \frac{q_k q_k^T}{\alpha_k p_k^T q_k} + \frac{g_k g_k^T}{p_k^T g_k} \tag{4.3}$$

where $\quad g_k^T = f^o_x(x_k) - \mu_k - \lambda_k f_x(x_k) = -p_k . H_k$ , (4.4)

and of course $\quad (x_k + p_k, \mu_k, \lambda_k) \in Z_r(x_k, H_k)$.

However, in this section we shall prove a result on the final convergence rate for the general class of "least-change" quasi-Newton formulae. The approach is essentially that of Dennis and Moré (16), using the results of Dennis and Walker (7, 8), but the treatment differs from theirs in detail.

Dennis and Walker consider the possibility of computing part of $H_{k+1}$ and approximating the remainder by a quasi-Newton formula, but although the development below could be readily generalized in this way, it is not likely to be useful for nonlinear programming and we have accordingly kept to the simpler treatment. We therefore assume the $H_k \in A \subset R^{n\times n}$, where $A$ is an affine subset, with parallel subspace $S$, which reflects the desired structure of the $H_k$. In the present context, $A$ will normally be the set of symmetric matrices, but could be $R^{n\times n}$ itself or further restricted to a specified sparsity pattern. We should also like to choose $H_{k+1}$ so that $H_{k+1}(\alpha_k p_k) = q_k$, but since this may not always be possible we define a "least-change secant formula" by the prescription:

$$H_{k+1} = \underset{H \in M_{W_k}(\alpha_k p_k, q_k)}{\arg\min} \| H - H_k \|_{W_k}, \tag{4.5}$$

where the set $M_W(p,q) \subset A$ is the set of matrices in $A$ closest to the set $Q(p,q) = \{M \in R^{nxn} | Mp = q\}$, using as a measure of "closeness" the weighted Frobenius norm $\| \cdot \|_W$. Thus $H_{k+1}$ is the matrix in $M_{W_k}(\alpha_k p_k, q_k)$ closest to $H_k$, using $\| \cdot \|_{W_k}$ as the measure of closeness.

The norm $\| \cdot \|_W$ is defined in terms of a given symmetric positive-definite matrix $W = J^2 \in R^{nxn}$ by

$$\| M \|_W = (tr(W^{-1} M W^{-1} M^T))^{\frac{1}{2}} = \| J^{-1} M J^{-1} \|_F, \; M \in R^{nxn} \tag{4.6}$$

where $\| \cdot \|_F$ is the Frobenius norm (given by (4.6) with W=I).

Similarly a "least-change inverse-secant formula" is defined by

$$S_{k+1} = \underset{S \in M_{W_k}(q_k, \alpha_k p_k)}{\arg\min} \| S - S_k \|_{W_k}, \tag{4.7}$$

where $S_{k+1}$ now approximates the inverse of the Hessian matrix, and $A$ reflects the latter's structure.

As indicated in (4.5) and (4.7), $W_k$ may change with the iteration, but we shall assume that

$$W_k \in W(\rho,\sigma) \triangleq \{W \in R^{nxn} | W = W^T \; ; \; \rho \| p \|^2 \leq p.W.p \leq \sigma \| p \|^2, \; p \in R^n\}, \tag{4.8}$$

and

$$\| W_{k+1} - W_k \| \leq K \| p_k \|^{\nu}, \tag{4.9}$$

for some constants $\sigma > \rho > 0$, $\nu > 0$, $K < \infty$, where $\| \cdot \|$ is the Euclidean norm.

We shall also need the linear subspace $\eta(p) = \{M \in R^{nxn} | Mp = 0\}$, and define $P_{A,W}$ as the orthogonal projection operator on the affine manifold $A$ in the norm $\| \cdot \|_W$ with $P^{\perp}_{A,W} = I - P_{A,W}$.

The basic result from Dennis and Walker (8) is then:

<u>Lemma 1</u>

If $p_k \neq 0$ and $H_{k+1}$ is given by (4.5), then for any $H_k \in R^{nxn}$, any $M \in A$, and any symmetric positive-definite $W \in R^{nxn}$ we have

$$\| H_{k+1} - M \|_W \leq \| P_{S \cap \eta(p_k), W_k}(H_k - M) \|_W + \| \bar{P}_k \frac{(q_k - M p_k) p_k^T}{p_k^T p_k} \|_W \tag{4.10}$$

where $\bar{P}_k = (I-P_{S,w_k}P_{\eta(P_k),w_k})^{-1}P_{S,w_k}P^{\perp}_{\eta(P_k),w_k}$.

Proof

The result follows by combining (2.24) and (3.10) of Dennis and Walker (8). They also show that $\bar{P}_k$ is bounded in norm uniformly in $p_k \varepsilon R^n$ when $A=R^{nxn}$ or $A$ is the set of symmetric matrices in $R^{nxn}$.

We shall also need the following:

Lemma 2

If $W,\bar{W}\varepsilon \mathcal{W}(\rho,\sigma)$, then there exists a scalar $\beta>0$, depending only on $\rho$ and $\sigma$, such that for any $M\varepsilon R^{nxn}$:

$$(1+\beta\| W-\bar{W}\|)^{-1}.\| M\|_W \leqslant \| M\|_{\bar{W}} \leqslant (1+\beta\| W-\bar{W}\|).\| M\|_W. \tag{4.11}$$

Proof

Writing $W=J^2$, $\bar{W}=\bar{J}^2$, we have from (4.5):

$$\begin{aligned}\| M\|_{\bar{W}} &= \| \bar{J}^{-1}M\bar{J}^{-1}\|_F \\ &= \| \bar{J}^{-1}JJ^{-1}MJ^{-1}J\bar{J}^{-1}\|_F \\ &= \| J\bar{W}J.J^{-1}MJ^{-1}\|_F \\ &\leqslant (1+\| J\bar{W}J-I\|).\| M\|_W \\ &\leqslant (1+\| J\bar{W}^{-1}(W-\bar{W})J^{-1}\|).\| M\|_W \\ &\leqslant (1+(\sigma/\rho^3)^{\frac{1}{2}}\| W-\bar{W}\|).\| M\|_W.\end{aligned}$$

The result follows by symmetry.

Q.E.D.

Finally, we need the following result, obtained from Theorems 11 and 12 in Sargent (12):

Lemma 3

Suppose that $Q_0(\hat{x},\hat{H})$ is strictly regular, with solution $\hat{x}$, and for each $\hat{z}\varepsilon Z_0(\hat{x},\hat{H})$ and each $u\varepsilon C_L'(\hat{z})$, the extended linearizing cone at $\hat{z}$, we have either $u.\hat{L}_{xx}(\hat{z}).u>0$ or $\hat{\mu}.u>0$ and $\hat{\lambda}.f_x(\hat{x}).u>0$. Then there exist $\gamma>0$ and open neighbourhoods $U(\hat{x})\subset R^n$, $V(\hat{H})\subset R^{nxn}$, such that for each $x\varepsilon U(\hat{x})$, $H\varepsilon V(\hat{H})$, $Q_0(x,H)$ is strictly regular and each $z\equiv(x+p,\mu,\lambda)\varepsilon Z_0(x,H)$ satisfies

$$d(z,Z_0(\hat{x},\hat{H}))\leqslant\gamma\{\| f_x^0(\hat{x}) + p\hat{H}-\mu-\lambda f_x(\hat{x})\| , \Delta x(\hat{x},p), \Delta f(\hat{x},p)\}, \tag{4.12}$$

where $\Delta x(\hat{x},p)$, $\Delta f(\hat{x},p)$ represent norms of the violations of the constraints (2.5) at $(\hat{x},p)$, and $d(z,Z) \triangleq \inf\{\| z-z'\| \,|z'\varepsilon Z\}$. Further, there exist $\gamma'>0$ and an open neighbourhood $W(z_0(\hat{x},\hat{H}))\subset R^nxR^nxR^m$ such that each $z\varepsilon Z_0(x,H)\cap W(Z_0(\hat{x},\hat{H}))$ satisfies

$$d(z,Z_0(\hat{x},\hat{H}))\leqslant\gamma'\{\| L_x(x+p,\mu,\lambda)\| , \Delta x(x+p), \Delta f(x+p)\}, \tag{4.13}$$

where $\Delta x(x+p)$, $\Delta f(x+p)$ represent norms of the violations of constraints (2.1) at $(x+p)$ and $L(x,\mu,\lambda)$ is the Lagrangian function for problem (2.1):

$$L(x,\mu,\lambda) = f^0(x) - \mu.\Delta\, x(x) - \lambda.\Delta\, f(x). \tag{4.14}$$

The extended linearizing cone $C_L'(\hat{z})$ is defined by

$$C_L'(\hat{z}) = \{u\varepsilon R^n | \hat{\mu}^i u^i \geqslant 0,\ i=1,2,\ldots n,\ \hat{\lambda}^j.f_x^j(\hat{x}).u \geqslant 0,\ j=1,2,\ldots m\}. \tag{4.15}$$

We now consider the asymptotic behaviour of the algorithm described in Section 3, with steps generated by (3.1) and the sequence $\{H_k\}$ generated by (4.5) with $\{W_k\}$ satisfying (4.8) and (4.9). For the present we assume $\alpha_k=1$, all k, but later we shall show that this is asymptotically true when $\alpha_k$ is chosen to satisfy (3.10) as described in Section 3.

Theorem 2

Suppose that $Q_0(\hat{x},\hat{H})$ is as in Lemma 3, and that $f^0(x)$, $f(x)$ are twice continuously differentiable on $U(\hat{x})$, with $f^o_{xx}(x)$, $f_{xx}(x)$ satisfying a Hölder condition of the form:

$$\| f_{xx}(x) - f_{xx}(\hat{x})\| < K\| x-\hat{x}\|^{\nu},\ x\varepsilon U(\hat{x}). \tag{4.16}$$

Suppose also that $A$ is such that the $\bar{P}_k$ in Lemma 1 are uniformly bounded for $x_k\varepsilon U(\hat{x})$, and that $\hat{H}\varepsilon A$ is invertible and satisfies

$$\| (\hat{H}-L_{xx}(\hat{x},\hat{\mu},\hat{\lambda}))p_k\| \leqslant K.\Delta_k^{\nu}.\| p_k\| , \tag{4.17}$$

where $\Delta_k = \max\{\| x_k-\hat{x}\| ,\ \| x_{k+1}-\hat{x}\| \}$.

Then there exist constants $\delta>0$, $\varepsilon>0$ such that if $\| x_0-\hat{x}\| \leqslant\delta$, $\| H_o-\hat{H}\| \leqslant\varepsilon$, the sequences $\{x_k\}$, $\{H_k\}$ are well defined by the above algorithm, $\{\| H_k\| \}$ and $\{\| H_k^{-1}\| \}$ are uniformly bounded, and the sequence $\{x_k\}$ remains in the closed ball $\bar{B}(\hat{x},\delta)\ \underline{\Delta}\ \{x\varepsilon R^n |\ \|x-\hat{x}\| \leqslant\delta\}\subset U(\hat{x})$ and converges Q-superlinearly to $\hat{x}$. Moreover, for each k and some $\hat{z}_k\varepsilon Z_0(\hat{x},\hat{H})$ we have

$$\| q_k-H_k p_k\| = o(\| p_k\| )\ ,\ \| \mu_k-\hat{\mu}_k\| = o(\| p_k\| ),\ \| \lambda_k-\hat{\lambda}_k\| = o(\| p_k\| ). \tag{4.18}$$

Proof

We can choose $\delta,\varepsilon$ so that $\bar{B}(\hat{x},\delta)\subset U(\hat{x})$, $\bar{B}(\hat{H},2\varepsilon)\subset V(\hat{H})$. It then follows (ref.12, Lemma 3) that $Z_0(x,H)$, $x\varepsilon\bar{B}(\hat{x},\delta)$, $H\varepsilon\bar{B}(\hat{H},2\varepsilon)$ is uniformly bounded, and we may therefore assume that $K,\nu$ are chosen so that (4.16) also applies to $L_{xx}(x,\mu,\lambda)$ for any fixed $(\mu,\lambda)$ corresponding to such $(x,H)$. It also follows that $L_{xx}(x,\mu,\lambda)$ and derivatives of $f^0(x)$, $f(x)$ are uniformly bounded. Thus (4.12) yields

$$d(z,Z_0(\hat{x},\hat{H})) \leqslant \gamma_1\| x-\hat{x}\| + \gamma_2\| H-\hat{H}\| \tag{4.19}$$

for some constants $\gamma_1>0$, $\gamma_2>0$. We can therefore choose $\delta,\varepsilon$ so that $Z_0(x,H)\subset W(Z_0(\hat{x},\hat{H}))$ for each $x\varepsilon\bar{B}(\hat{x},\delta)$, $H\varepsilon\bar{B}(\hat{H},2\varepsilon)$, and hence from (4.13) we have

$$d(z,Z_0(\hat{x},\hat{H})) \leqslant \gamma\| L_x(z) + p.L_{xx}(z)\| + o(\| p\| ). \tag{4.20}$$

Now suppose that for some $k\geqslant 0$, and some $r\varepsilon(0,1)$:

$$H_j\varepsilon\bar{B}(\hat{H},2\varepsilon)\ ,\ \| x_j-\hat{x}\|\leqslant r\,\|x_{j-1}-\hat{x}\| \ ,\ j=1,2,\ldots k. \tag{4.21}$$

It follows that each $Q_0(x_j,H_j)$ is strictly regular. Further, since $\hat{H}$ is invertible, we can choose $\varepsilon$ so that $\| H_k-\hat{H}\| .\| \hat{H}^{-1}\| \leqslant \kappa < 1$, and hence by the Banach perturbation lemma (ref. 15), $H_k^{-1}$ exists and $\| H_k^{-1}\| \leqslant \| \hat{H}^{-1}\| /(1-\kappa)$. It follows that (4.4) has a solution and $x_{k+1}$ is well defined.

Now from (4.2), (4.14), (4.16) and the mean-value theorem, for any $\hat{z}\varepsilon Z_0(\hat{x},\hat{H})$ we have

$$\| L_{xx}(z_k) - L_{xx}(\hat{z})\| \leqslant K\Delta_k^{\nu} + K_2\| \lambda_k-\hat{\lambda}\| , \tag{4.22}$$

and $\| q_k - L_{xx}(\hat{z}).p_k \| \leqslant \| p_k \| (K.\Delta_k^{\nu} + K_2 \| \lambda_k - \hat{\lambda} \| )$ (4.23)
where $K_2$ is the bound on $\| f_{xx}(x) \|$.

Thus from (4.4), (4.17), (4.19), (4.20) and (4.22):

$$\begin{aligned} d(z_k, Z_0(\hat{x},\hat{H})) &\leqslant \gamma \| (H_k - L_{\hat{x}x}(z_k))p_k \| + o(\| p_k \|) \\ &\leqslant \gamma \| (H_k - \hat{H})p_k \| + o(\| p_k \|), \end{aligned} \tag{4.24}$$

and since $x_{k+1} = x_k + p_k$ and $\| p_k \| \leqslant 2\Delta_k$, it follows that we may choose $\delta$ and $\varepsilon$ so that $\| x_{k+1} - \hat{x} \| \leqslant r \| x_k - \hat{x} \|$.

Now from the hypotheses on $A$ and $\{W_k\}$, and using (4.17) and (4.23), we have for the second term in (4.10), with $M=\hat{H}$, $W=W_k$, and for some $\beta'$:

$$\| \bar{P}_k \frac{(q_k - \hat{H}p_k)p_k^T}{p_k^T p_k} \|_{W_k} \leqslant \beta' \frac{\| q_k - \hat{H}p_k \|}{\| p_k \|} \leqslant 2\beta' K.\Delta_k^{\nu} + \beta' K_2 \| \lambda_k - \hat{\lambda} \| . \tag{4.25}$$

For the first term in (4.10) we have

$$\begin{aligned} \| P_{S \cap \eta(p_k), W_k} (H_k - \hat{H}) \|_{W_k} &\leqslant \| P_{S \cap \eta(P_k), W_k} . P_{\eta(P_k), W_k} (H_k - \hat{H}) \|_{W_k} \\ &\leqslant \| (H_k - \hat{H}) \left( I - \frac{P_k P_k^T W_k}{P_k^T W_k P_k} \right) \|_{W_k} \\ &\leqslant (1-\theta_k^2)^{\frac{1}{2}} \| H_k - \hat{H} \|_{W_k} \end{aligned} \tag{4.26}$$

where $\theta_k = 0$ if $H_k = \hat{H}$ and otherwise

$$\theta_k = \frac{\| (H_k - \hat{H})p_k \|}{\| W_k \| . \| H_k - \hat{H} \|_{W_k} . \| P_k \|} < 1 \tag{4.27}$$

Thus from (4.10), (4.25) and (4.26), and then using Lemma 2 and (4.9), we have the "bounded deterioration" property:

$$\| H_{k+1} - \hat{H} \|_{W_{k+1}} \leqslant (1 + \beta_1 \| P_k \|^{\nu}).(1-\theta_k^2)^{\frac{1}{2}}. \| H_k - \hat{H} \|_{W_k} + \beta_2 \Delta_k^{\nu} + \beta_3 \| \lambda_k - \hat{\lambda} \| , \tag{4.28}$$

for some constants $\beta_1$, $\beta_2$, $\beta_3$. Thus, using (4.24):

$$\| H_{k+1} - \hat{H} \|_{W_{k+1}} \leqslant \| H_o - \hat{H} \|_{W_o} + \sum_{j=o}^{k} (2\varepsilon\beta_1 \| P_k \|^{\nu} + \beta_2 \Delta_k^{\nu} + 2\varepsilon\beta_3\gamma \| P_k \| + o(\| P_k \| )), \tag{4.29}$$

and in view of (4.21) we may further choose $\delta, \varepsilon$ so that $H_{k+1} \varepsilon \bar{B}(\hat{H}, 2\varepsilon)$.
But clearly these arguments hold for k=0, so by induction (4.21) holds for all $k \geqslant 0$.

Now, rearranging (4.28) we have

$$(1-(1-\theta_k^2)^{\frac{1}{2}}) \| H_k - \hat{H} \|_{W_k} \leqslant \| H_k - \hat{H} \|_{W_k} - \| H_{k+1} - \hat{H} \|_{W_{k+1}} + (2\varepsilon\beta_1 \| P_k \|^{\nu} + \beta_2 \Delta_k^{\nu} + 2\varepsilon\gamma \| p_k \| + o(\| p_k \|)),$$

and summing both sides:

$$\sum_{k=o}^{\infty} (1-(1-\theta_k^2)^{\frac{1}{2}}. \| H_k - \hat{H} \|_{W_k} < \infty. \tag{4.30}$$

Now from Dennis and Moré (16, Lemma 3), (4.29) implies that $\{ \| H_k - \hat{H} \|_{W_k} \}$ converges, so if some subsequence of $\{ \| H_k - \hat{H} \|_{W_k} \}$ converges to zero, then the whole sequence converges to zero. Otherwise this sequence is bounded away from zero, and then from (4.30) $\{\theta_k\} \to 0$. In both cases we therefore have

$$\lim_{k\to\infty} \frac{\| (H_k-\hat{H})P_k \|}{\| P_k \|} = 0 \tag{4.31}$$

Conditions (4.18) and the Q-superlinear convergence of $\{x_k\}$ then follow from (4.24) and (4.25).

Q.E.D.

Condition (4.17) requires some discussion. Clearly, if $L_{xx}(\hat{z})$ is non-singular we may simply take $\hat{H}=L_{xx}(\hat{z})$, and (17) is satisfied with no restrictions on $P_k$. More generally, we may choose $\hat{H}$ so that $\hat{H}u=L_{xx}(\hat{z}).u$, $u\varepsilon\bar{C}_L(\hat{x})=\cup\{C_L'(\hat{z})|z\varepsilon Z_0(\hat{x},\hat{H})\}$, but then $P_k$ must be close to $\bar{C}_L(\hat{x})$. In this connection, we note from (2.5) that

$$\lambda_k(f(x_{k+1})-c)=\lambda_k(f(x_k)+f_x(x_k).P_k+o(\| P_k\| )-c) = o(\| P_k\| ). \tag{4.32}$$

Hence, using (4.18):

$$\lambda_k f_x(x_k).P_k = -\lambda_k(f(x_k)-c) = (\lambda_{k-1}-\lambda_k)(f(x_k)-c)-\lambda_{k-1}(f(x_k)-c) = o(\| P_k\| ),$$

and similarly for other bounds.

Thus by choice of $\delta$ we can in fact make $P_k$ arbitrarily close to $\bar{C}_L(\hat{x})$ for all $k$ and $\lim_{k\to\infty}P_k/\| P_k\| \;\varepsilon\; \bar{C}_L(\hat{x})$.

For the choice of $\alpha_k$ using (3.10), we have from (4.14):

$$L(x_k,\mu_k,\lambda_k) - L(x_{k+1},\mu_{k+1},\lambda_{k+1}) = -L_x(x_k,\mu_k,\lambda_k).P_k-\tfrac{1}{2}P_k.L_{xx}(x_k,\mu_k,\lambda_k).P_k$$
$$+ (\mu_{k+1}-\mu_k)\Delta x(x_{k+1}) + (\lambda_{k+1}-\lambda_k)\Delta f(x_{k+1}) + o(\| P_k\| ^2),$$

and using (4.4), (4.17), (4.18), (4.22), (4.30) and (4.31) this yields

$$L(x_k,\mu_k,\lambda_k) - L(x_{k+1},\mu_{k+1},\lambda_{k+1}) = \tfrac{1}{2}P_kH_kP_k + o(\| P_k\| ^2).$$

Further, from (4.18) and (4.31) we have $(\lambda_{k+1}-\lambda_k)\Delta f(x_{k+1}) = o(\| P_k\| ^2)$. Thus, if in (3.10) we set $\delta\varepsilon(0,\frac{1}{2})$, the $\delta$ of Theorem 2 can be chosen small enough for (3.10) to be satisfied with $\alpha_k=1$, and (3.9) yields $\bar{\alpha}_{k+1} = 1$, for all $k\geqslant 0$. Finally, from (4.18): $P_kH_kP_k = P_k\hat{H}P_k + o(\| P_k\| ^2)$, and in view of the second-order sufficiency condition $u.\hat{H}.u \geqslant \rho\| u \| ^2$, $u\varepsilon\bar{C}_L(\hat{x})$, for some $\rho>0$. Thus $\delta$ can be chosen so that $\{H_k\}$ satisfies the conditions of Theorem 1.

If the DFP or BFGS formula is used to generate $\{H_k\}$, it is well known that all $H_k$ are positive-definite if $H_o$ is positive-definite and $P_k^Tq_k>0$ for each $k$. In case of failure of the latter condition, we may simply set $H_{k+1} = H_k$, and then the conditions of Theorem 1 are satisfied for any finite $k$. Should this yield a point satisfying the conditions of Theorem 2, final convergence would be at a Q-superlinear rate, but there is a gap between the two results since we may never have $\| H_k-\hat{H}\|$ sufficiently small.

Nevertheless, Theorem 2 is an advance on previously published results. The most general earlier result is that of Josephy (10), using Robinson's (17) results on "generalized equations". This relaxed the requirement of "strict complementarity" used by earlier workers, but still needed linear independence of the active constraint gradients, as well as the extended second-order sufficiency condition used in Theorem 2. Finally, Theorem 2 can also be proved for least-change inverse-secant methods; it also holds (except for the existence of $H_k^{-1}$) for the Newton choice $H_k = L_{xx}(z_k)$.

## 5. Feasible-Point Algorithms

In many cases the objective function and constraint functions behave badly, or may even be undefined, outside the feasible region. It is usually possible to add constraints which exclude regions where the function is undefined, and if the initial point is physically reasonable, even though unfeasible, it is usually in a region where the functions are reasonably behaved. For such situations it is therefore desirable to have an algorithm which first concentrates on obtaining a feasible point, and once such a point is attained maintains feasibility in subsequent iterations.

It turns out that the recursive quadratic programming algorithm is easily modified to achieve this, for if the current point $x_k$ is feasible the solution $P_k$ defines a feasible direction, and otherwise defines a descent direction for reduction of the constraint violations:

From continuity of $f_x(x)$ we have

$$f(x+\alpha p) = f(x) + \alpha f_x(x).p + o(\|\alpha p\|) \tag{5.1}$$

and using (2.6), (2.7) and (2.8), with $\alpha\varepsilon(0.1]$:

$$\begin{aligned} f^j(x+\alpha p) &\geqslant f(x) + \alpha\bar{C}^j_r + o(\|\alpha p\|) \\ &\geqslant C^j - \bar{C}^j_o + \alpha(\bar{C}^j_o - \Delta C^j_o + \Delta C^j_r) + o(\|\alpha p\|) \\ &\geqslant C^j - \Delta C^j_o + \alpha\Delta C^j_r + o(\|\alpha p\|) \end{aligned} \tag{5.2}$$

Thus, since $\Delta C^j_r>0$, for any $\delta\varepsilon(0,1)$ there is an $\bar{\alpha}\varepsilon(0,1]$ such that for each $\alpha\varepsilon(0,\bar{\alpha}]$ we have

$$f^j(x+\alpha p) \geqslant C^j - \Delta C^j_o + \delta.\alpha\Delta C^j_r = C^j - (1-\delta\alpha\theta^r)\Delta C^j_o. \tag{5.3}$$

Now if $f^j(x) \geqslant C^j - \varepsilon$ for any $j\varepsilon NL$, we have $\Delta C^j_o = \varepsilon$ and hence $f^j(x+\alpha p) > c^j - \varepsilon$, so that feasibility is maintained. On the other hand, if $f^j(x) < c^j - \varepsilon$ we have $\Delta C^j_o = \bar{C}^j_o$, and from (5.3):

$$c^j - f^j(x+\alpha p) \leqslant (1-\delta\alpha\theta^r)(c^j - f^j(x)), \tag{5.4}$$

so that the constraint violation is reduced by the factor $(1-\alpha\delta\theta^r)<1$. A similar argument holds for the upper bound.

Thus it is possible to use the Armijo rule to reduce $\alpha_k$ until (5.4) is satisfied for all violated constraints and feasibility is maintained for other nonlinear constraints. Once a feasible point is attained, $\alpha_k$ is reduced to maintain feasibility of the nonlinear constraints while at the same time satisfying (3.10) for the Lagrangian descent function.

For this algorithm, convergence to an initial feasible point is linear, whilst Theorems 1 and 2 apply to the second phase in which feasibility is maintained, since when $\| P_k \|$ becomes sufficiently small (5.2) shows that feasibility is maintained with $\alpha_k = 1$.

Of course, a new quadratic programme must be solved for each $\alpha$ in seeking to satisfy (3.10), and although the successive solutions are quickly obtained since each is only a small perturbation of the previous one, successive evaluation of constraint functions and gradients is required. This work can be avoided by using the objective function itself as a descent function.

Thus if x is feasible, for a nonlinear constraint we have $c^j-\varepsilon\leq f^j(x)\leq d^j+\varepsilon$ and $\lambda^j(c^j-\varepsilon-f^j(x)-f_x^j(x).p)\geq 0$, $\lambda^j(d^j+\varepsilon-f^j(x)-f_x^j(x).p)\geq 0$, from which it follows that $\lambda^j f_x^j(x).p\leq 0$. With similar reasoning for other constraints, we have from (2.4):

$$f_x^o(x).p + p.H.p = \mu.p + \lambda.f_x(x).p \leq 0,$$

and if H is positive definite:

$$-f_x^o p \geq p.H.p > 0. \tag{5.5}$$

Hence, expanding $f^0(x)$ as in (5.1) and using (5.5), we see that

$$f^0(x_k) - f^0(x_k+\alpha_k p_k) > \delta\alpha_k p_k.H_k p_k > 0, \tag{5.6}$$

for any $\delta\varepsilon(0,1)$ and all $\alpha_k$ sufficiently small. We note however that we do not obtain $\alpha_k=1$ when $\alpha_k$ is sufficiently small, since the second-order terms become $\frac{1}{2}p_k(H_k-f_{xx}^o(x_k).p_k$, which remain of the same order as $p_k H_k p_k$. Thus the ultimate convergence rate is only linear. Nevertheless, the zone in which a superlinear rate would be effective using $L(r,\alpha,z)$ may be very small, and the saving in computation per iteration could well more than compensate for the slower ultimate convergence rate.

Whatever descent function is used in the second phase, it is well known that feasible-point algorithms of this type can be very slow in following a curved constraint, and even for the first published algorithm Rosen (18) suggested the use of a Newton-type correction procedure for satisfying such constraints. Sargent (11) reformulated this procedure as a subsequence of QP problems, which in the present context may be written:

$$\left.\begin{array}{lr} \text{Minimize} & \bar{g}_{t-1}^T\ \Delta p_t + \tfrac{1}{2}\Delta p_t^T H\Delta p_t \\ \text{subject to} & a\leq x_{t-1} + \Delta p_t\leq b \\ & c^j\leq f^j(x_{t-1}) + f_x^j(x).\Delta p_t\leq d^j\ \ ,\ j\notin NL, \\ & \alpha c^j\leq f^j(x_{t-1})+f_x^j(x).\Delta p_t-(1-\alpha)f^j(x)\leq\alpha d^j\ \ ,\ j\varepsilon NL. \end{array}\right\} \tag{5.7}$$

where

$$x_0\equiv x,\ x_t=x_{t-1}+\Delta p_t\ ,\ \bar{g}_t^T=\alpha f_x^o(x)+p_t.H \tag{5.8}$$

The scaling by $\alpha$ here has an effect similar to the relaxation procedure described earlier, and the latter should not therefore be necessary with this version of the algorithm. If $x\varepsilon X$, the initial solution $p_1$ can be made as small as desired by choosing $\alpha$ sufficiently small, and since the iterations of the subsequence form a simplified Newton procedure (without re-evaluation of gradients) convergence is assured for $\alpha$ sufficiently small, and the convergence rate is linear. A suitable termination test for the iterations is

$$|f_x(x).\Delta p_t| \leq \delta'\alpha.\max(\varepsilon,c^j-f^j(x),\ f^j(x)-d^j),\ j\varepsilon NL \tag{5.9}$$

It is easily verified that this maintains feasibility of feasible constraints, and reduces violations by the factor $(1-\alpha(1-\delta'))<1$, so the correction procedure can be used in both phases of the programme. In practice, the Armijo rule is used to reduce $\alpha_k$ until the correction procedure satisfies (5.9) within a fixed number of steps (e.g. 10 steps), and of course in the second phase until the descent test is also satisfied within this number.

Although this correction procedure re-introduces the need to solve a QP subproblem for each t, only re-evaluations of the constraint functions are required. Each solution is of course an excellent starting point for the next subproblem, and the solution is particularly efficient if, as suggested by Sargent (11), the Lemke algorithm (for solving the equivalent linear complementarity problem) is used, with the "re-start" procedure described by Byrne (19).

It is perhaps worth noting that the Han version of recursive quadratic programming, with automatic increase of the penalty function weighting as advocated by Mayne and Polak (5), could well behave like the first feasible-point method, without correction procedure, described above. This is because a high penalty weight gives emphasis to attaining feasibility, followed by small steps along curved constraints to avoid later large violations. The weighting in our own descent function is not under the control of the user, but it is hoped that it provides the "natural" balance between constraint violation and objective function decrease. Nevertheless, the choice between relying on this descent function alone or using the feasible-point method with the correction procedure is not at all obvious, and can only be clarified by numerical testing.

References

1. Sargent, R.W.H., "A Review of Optimization Methods for Nonlinear Problems", in R.G. Squires and G.V. Reklaitis (Eds) "Computer Applications to Chemical Engineering", ACS Symposium Series 124, 37-52, 1980.
2. Wilson, R.B., "A Simplicial Method for Convex Programming", PhD Dissertation, Harvard University, 1963.
3. Han, S.P., "Superlinearly Convergent Variable-Metric Algorithms for General Nonlinear Programming Problems", Math. Programming, 11(3), 236-282, (1976).
4. Han, S.P., "A Globally Convergent Method for Nonlinear Programming", JOTA, 22(3), 297-309, (1977).
5. Mayne, D.Q., "A Superlinearly Convergent Algorithm for Constrained Optimization Problems, in G.A. Watson (Ed) "Numerical Analysis Proceedings, Dundee 1979", Lecture Notes in Mathematics, 773, 98-109, Springer-Verlag, Berlin, 1980.
6. Chamberlain, R.M., C. Lemarechal, H.C. Pedersen, and M.J.D. Powell, "The Watchdog technique for Forcing Convergence in Algorithms for Constrained Optimization", presented at Tenth International Symposium on Mathematical Programming, Montreal, August 1979.
7. Dennis, Jr., J.E., and H.F. Walker, "Local Convergence Theorems for Quasi-Newton Methods", Rice University Report No. TR476-(141-171-163)-1, Houston, 1980.
8. Dennis, Jr., J.E., and H.F. Walker, "Convergence Theorems for Least-Change Secant Update Methods", Rice University Report No. TR476-(141-171-163)-2, Houston, 1980.
9. Powell, M.J.D., "The Convergence of Variable Metric Methods for Nonlinearly Constrained Optimization Calculations", in O.L. Mangasarian, R.R. Meyer and S.M.

Robinson (Eds), "Nonlinear Programming 3", 27-64, Academic Press, New York 1978.

10. Josephy, N.H., "Quasi-Newton Methods for Generalized Equations", MRC Report No. 1966, University of Wisconsin-Madison, June 1979.
11. Sargent, R.W.H., "Reduced Gradient and Projection Methods for Nonlinear Programming", in P.E. Gill and W. Murray (Eds) "Numerical Methods for Constrained Optimization", Academic Press, London 1974.
12. Sargent, R.W.H., "On the Parametric Variation of Constraint Sets and Solutions of Minimization Problems", presented at Tenth International Symposium on Mathematrical programming, Montreal, August 1979. To appear in Mathematical Programming Study 15.
13. Powell, M.J.D., "A Fast Algorithm for Nonlinearly Constrained Optimization Calculations", in G.A. Watson (Ed) "Numerical Analysis - Proceedings of Biennial Conference, Dundee 1977", Lecture Notes in Mathematics, 630, 144-159, Springer-Verlag, Berlin 1978.
14. Polak, E., R.W.H. Sargent and D.J. Sebastian, "On the Convergence of Sequential Minimization Algorithms", JOTA, 14(4), 439-442, (1974).
15. Ortega, J.M., and W.C. Rheinboldt, "Iterative Solution of Nonlinear Equations in Several Variables", Academic Press, New York, 1970.
16. Dennis, Jr., J.E., and J.J. Moré, "A Characterization of Superlinear Convergence and its Application to Quasi-Newton Methods", Math.Comp. 28(126), 549-560,(1974).
17. Robinson, S.M., "Strongly Regular Generalized Equations", Maths of Operations Research, 5(1), 43-62, (1980).
18. Rosen, J.B., "The Gradient Projection Method for Nonlinear Programming: Part II: Nonlinear Constraints", J.SIAM, 9(4), 514-532, (1961).
19. Byrne, S.J., "Quadratic Programming Using Complementarity and Orthogonal Factorization", PhD Thesis, University of London, 1980.

# Dual and Primal-Dual Methods for Solving Strictly Convex Quadratic Programs

by

D. Goldfarb

The City College of New York, CUNY, New York, N.Y. 10031

and

A. Idnani

Bell Laboratories, Murray Hill, N.J. 07974

## 1. Introduction

In this paper we consider the strictly convex quadratic programming (QP) problem

$$\underset{x}{\text{minimize}}\ f(x) = f_0 + a^T x + 1/2\, x^T G x \qquad (1a)$$

$$\text{subject to} \quad s(x) \equiv C^T x - b \geqslant 0, \qquad (1b)$$

where $x$ and a are n-vectors, G is an nxn symmetric positive definite matrix, C is an nxm matrix, b and the slack vector $s(x)$ are m-vectors, and superscript T denotes transposition. The development of an algorithm to solve such problems efficiently is currently of some importance as many nonlinear programming methods require the solution of a strictly convex quadratic program to determine a search direction at each iteration (e.g., see [1,7,8,10]).

If the usual primal approach is adopted for solving problem (1), then unless an initial feasible point is available, some phase 1 procedure is required. Unfortunately, such phase 1 calculations often use up as much or nearly as much time as it takes the primal algorithm to obtain an optimal solution from the feasible solution provided by the phase 1 procedure. On the other hand, dual methods for solving problem (1) do not require an initial primal feasible point. Moreover, they can always be started from the unconstrained minimum $x_0=-G^{-1}a$ of (1a) so that no phase 1 is needed. From the point of view of the quadratic program dual to (1), this corresponds to the fact that the origin in the space of dual variables is always feasible. For this reason, we believe that dual methods are superior to purely primal methods when a feasible point for problem (1) is not readily available.

In this paper we present an approach, first given in [9], for solving (1), which is based upon finding the optimal solutions to a sequence of subproblems of (1), each corresponding to a relaxation of some (initially all) of the constraints (1b). Each subproblem in the sequence, other than the first, is required to contain a constraint that is violated at the optimal solution of the previous subproblem. This approach leads quite naturally to two different algorithms one which is a purely dual algorithm and one which is a primal-dual algorithm.

In the next section we describe this approach and these two algorithms in terms of the primal problem (1). One could also provide a description in terms of a program dual to (1), but we believe that the description in terms of the primal is more instructive. An example is presented in section 3 to illustrate that although both algorithms have a common development they may follow different solution paths, finding the optimal solutions to different sequences of subproblems of (1). Finally in section 4 we report some computational results which support our claim that dual methods are superior to primal methods for solving strictly convex quadratic programs when a feasible point is not known.

We only outline here the basic algorithms. A more complete description of their implementation using numerically stable procedures for updating matrix factorizations is given in [6,9]. Proofs of the finite termination of both algorithms, their relationship to other published quadratic programming algorithms, and a fuller presentation of the computational results can also be found in [6,9].

## 2. Basic Approach and Algorithms

Our basic approach for solving problem (1) is based upon finding the optimal solutions of a sequence of subproblems of (1) with strictly increasing optimal objective function values. Initially we solve for the unconstrained minimum of the quadratic objective function (1a); that is, we start with the subproblem P(A) defined by (1a) subject to the subset of constraints in (1b) consisting of the empty set $A=\emptyset$. Henceforth we shall use A to denote both a subset of constraints and their indices. Then having the optimal solution of a particular subproblem P(A) we check if it satisfies all of the constraints (1b). If it does then it clearly is the optimal solution of the full problem (1). If it does not, then we choose some constraint, say the p-th, violated by the solution. If the constraints in $A \cup \{p\}$ are inconsistent, then (1b) must also be inconsistent (infeasible) and we stop. Otherwise, because of the strict convexity of f(x), we can always define a new subproblem $P(\bar{A} \cup \{p\})$, with $\bar{A} \subseteq A$, such that the optimal value of f(x) for it is greater than it is for P(A).

To do this we proceed as follows. Let us note that the set of constraints A, which define the current subproblem P(A), are <u>active</u> (i.e., satisfied as equalities) at its optimal solution x. We first solve for the point $\bar{x}$ which minimizes f(x) over the constraints in A U {p} treated as equalities. (If the normals to the constraints indexed by A U {p} are linearly dependent we must first drop one of the constraints from the active set A before doing this.) If $\bar{x}$ is also optimal for the subproblem P(A U {p}) where now, once again the constraints indexed by AU {p} are treated as inequalities, then $\bar{x}$ and P(A U {p}) become our new optimal solution and subproblem, and a major iteration is complete. If $\bar{x}$ is not an optimal solution of P(A U {p}) we give two different methods for proceeding.

In the first method we find the point $\hat{x}$ on the line segment $[x,\bar{x}]$ between x and $\bar{x}$ closest to $\bar{x}$ which is an optimal solution to the subproblem obtained from P(A U {p}) by relaxing (i.e., translating) the p-th constraint so that $\hat{x}$ satisfies it as an equality. This point $\hat{x}$ is where one of the dual variables (i.e. Lagrange multipliers), which vary linearly as we move from x to $\bar{x}$ along $[x,\bar{x}]$, first becomes zero and is about to change sign. The constraint (say the k-th) corresponding to this dual variable is then dropped from the active set A and we repeat the above procedure starting with $\hat{x}$ in place of x and $A\setminus\{k\}$ in place of A. Eventually, after at most $|A|$ repetitions of this procedure the optimal solution of the equality constrained problem determined by $\bar{A}$ U {p} where $\bar{A} \subseteq A$ is also optimal for $P(\bar{A} \cup \{p\})$. Since the dual variables always stay nonnegative in this procedure, it is clearly a dual (feasible) method.

In the second method, we simply define our new subproblem to be P(A U {p}) and determine its optimal solution by applying a primal algorithm to this subproblem starting from the feasible but nonoptimal point $\bar{x}$. This naturally results in a priumal-dual algorithm.

Before we formally present these two algorithms we need to introduce some notation and definitions. We denote by N the matrix of normal vectors to the constraints in the active set A. The cardinality of A, and hence the column dimension of N, will be denoted by q. Assuming that N has linearly independent columns we can define the operators

$$N^* = (N^TG^{-1}N)^{-1}N^TG^{-1} \qquad (2)$$

and

$$H = G^{-1}(I-NN^*). \qquad (3)$$

$N^*$ is the Moore-Penrose generalized inverse of N in the space of variables obtained under the transformation $y=G^{1/2}x$. H is a "reduced" inverse Hessian operator for ther quadratic function f(x) subject to the active set of constraints treated as equalities.

Fundamental to the development of our algorithms is the fact that points x and $\bar{x}$ discussed above are related by the equation

$$\bar{x} = x + tz, \quad \text{where} \quad z = Hn_p, \quad t = -s_p(x)z^Tn_p,$$

$n_p$ is the normal vector corresponding to constraint p, and $s_p(x)$ is the slack in that constraint at x. Moreover, it can be shown that if t is considered to be a variable step length in the direction z then the dual variable $u(x)=[N:n_p]*g(x)$ corresponding to the constraints indexed by $A \cup \{p\}$ vary linearly according to the relation

$$u(\bar{x}(t)) = u(x) + t\binom{-r}{1},$$

where $r = N^*n_p$. (See Lemma 1 in [6].)

Dual Algorithm:

0) Find the unconstrained minimum:

Set $x \leftarrow -G^{-1}.a$, $f \leftarrow 1/2a^Tx+f_0$, $H \leftarrow G^{-1}$, $A \leftarrow \emptyset$, $q \leftarrow 0$.

1) Choose a violated constraint, if any: (x is the optimal solution of P(A).)

Compute $s_j(x)$, for all $j \notin A$

If $V = \{j \notin A | s_j(x) < 0\} = \emptyset$, STOP; the current solution x is both feasible and optimal.

Else, choose $p \in V$ and set $u \leftarrow \binom{u}{0}$. (If q=0, set $u \leftarrow 0$).

2) Check for feasibility and determine the optimal solution of a new subproblem:

2a) Compute $z = Hn_p$ (the step direction),

and $r = N^*n_p$ (the directional derivative of u(x) along z).

If $r \leqslant 0$ (or $q = 0$) set $t_1 = \infty$ ; else set

$$t_1 = \min_{\substack{r_j > 0 \\ 1 \leqslant j \leqslant q}} \left\{ \frac{u_j(x)}{r_j} \right\} = \frac{u_l(x)}{r_l}$$

(Let the index of the constraint corresponding to this minimum be k.)

2b) If $|z| = 0$, (find a constraint to drop.)

If $t_1 = \infty$, STOP; subproblem $P(A \cup \{p\})$, and hence problem (1), are infeasible. Else, drop constraint k; i.e., set $A \leftarrow A \setminus \{k\}$, $q \leftarrow q-1$, $u \leftarrow u + t_1\binom{-r}{1}$, delete component l of u, update H and N*, and go to (2a).

2c) Else, (i.e. $|z|=\emptyset$) compute $t_2 = -s_p(x)/z^Tn_p$ and set $t \leftarrow \min\{t_1, t_2\}$,

$$x \leftarrow x + tz, \qquad f \leftarrow f + tz^Tn_p(t/2 + u_{q+1}) \text{ and } u \leftarrow u + t\binom{r}{1}$$

If $t_2 \leqslant t_1$ (full step), set $A \leftarrow A \cup \{p\}$, update H and N*, and go to (1).

Else, ($t_1 < t_2$, a partial step), set $A \leftarrow A \setminus \{k\}$, $q \leftarrow q - 1$, delete component l of u, update H and N*, and go to (2a).

Primal-Dual Algorithm.

Steps (0), (1), (2a) and (2b): Same as in the Dual algorithm.

Step (2c): Else, (i.e. $|z| \neq 0$) compute $t_2 = -s_p(x)/z^T n_p$ and set

$f \leftarrow f - s_p(x)(t/2 + u_{p+1})$, $x \leftarrow x + t_2 z$ and $u \leftarrow u + t_2 \begin{Bmatrix} -r \\ 1 \end{Bmatrix}$

If $u \geqslant 0$, set $A \leftarrow A \cup \ p$ , $q \leftarrow q+1$, update H and N*, and go to (1).

Else, solve $P(A \cup \{p\})$ by a primal algorithm computing

A, q, f, u, H, and N* at the optimal point x, and go to (1).

This algorithm is not fully specified without indicating which particular primal algorithm is used in step (2c). Many choices are possible. In the results of the computational tests reported in section 4 the primal algorithm used in step (2c) is the one referred to as algorithm 1 in [5].

One important advantage of the primal-dual algorithm is that it can be started from an arbitrary initial point $x^o$. To do this one determines the set of constraints A satisfied by $x^o$, solves P(A) by the primal algorithm, and then goes to step (1). One could, of course, also continue from step (1) with the dual algorithm. If this is done then the primal algorithm acts as a phase 1 procedure for computing a dual feasible starting point for the dual method. Several variants of the primal-dual algorithm are possible. One that seems to be promising is to apply the primal algorithm to P(S) where S is the set of <u>all</u> satisfied constraints at the point x determined by a full step in step (2c).

Also in our implementation of both of these algorithms the operators H and N* are not explicitly computed or updated as the active set changes. Rather we store and update the matrices $J = Q^T L^{-1}$ and R, where Q and R correspond to the QR factorization

$$L^{-1}N = Q\begin{bmatrix} R \\ 0 \end{bmatrix}$$

and L comes from the Cholesky factorization

$$G = LL^T.$$

Numerically stable procedures for updating J and R are described in [6,9].

## 3. An Example

In this section we illustrate both the dual and primal dual algorithms by applying them to a problem in three variables with three constraints. Clearly, the solution paths taken by these algorithms in going from the optimal solution of the current subproblem P(A) to the optimal solution of the new subproblem $P(\bar{A} \cup \{p\})$ will be different when $\bar{A} \neq A$, i.e., when $t \neq t_2$ in step (2c) in the dual algorithm. This is true even if the new subproblems defined by both algorithms are the same. Our principal purpose in presenting the example below is to show that the new subproblems determined by the two algorithms need not be the same.

Problem:

Minimize $f(x) = 1/2x_1^2 + 1/2(x_2-5)^2 + 1/2x_3^2$

subject to

$$\begin{aligned} -4x_1-3x_2 &\geq -8 \\ 2x_1+x_2 &\geq 2 \\ -2x_2+x_3 &\geq 0 \end{aligned}$$

This problem is depicted in Figure 1. We first give the computations performed by the dual algorithm in solving it.

Solution by Dual Algorithm

Initially set $x^O=(0,5,0)^T$, the unconstrained minimum of $f(x)$, $f=0$, $H=G^{-1}=I, A=\emptyset$ and $q=0$.

Iteration 1: Compute $s(x^O)=(-7,2,-5)^T$. Choose $p = 1$ and take a full step; i.e., compute

$$z=Hn_1=(-4,-3,0)^T,\ t=t_2=-s_2(x^O)/z^Tn_1=7/25$$

$$x^1=x^O+tz=\frac{1}{25}(-28,104,0)^T, f=49/50 \quad \text{and } u=7/25.$$

Note: $\nabla f(x) = u(x)n_1$ with $u(x)>0$ for all $x \in [x^O, x^1]$. Adding constraint 1 to the active set yields

$$A = \{1\}, q = 1, H = \frac{1}{25}\begin{bmatrix} 9 & -12 & 0 \\ -12 & 16 & 0 \\ 0 & 0 & 25 \end{bmatrix} \text{ and } N^* = \frac{1}{25}(-4,-3,0)$$

Iteration 2: $x^1$ is the optimal solution of $P(\{1\})$.

Compute $s(x^1)=(0,-2/25,-208/25)^T$.

Choose $p = 2$. Note that we can choose any violated constraint to add to the active set. Now calculate

$z = 1/25(6,-8,0)^T$, $z^Tn_2=4/25$, $r=-11/25$

$t_1=\infty$, and $t_2 = 1/2$. Since $t_2 < t_1$, set $t=t_2$ and take a full step; i.e., compute

$x^2 = x^1 + tz = (-1,4,0)$

$f = 49/50 + 1/2 \cdot 4/25\ (1/4+0) = 1$

and $u = (7/25,0)^T + 1/2(11/25,1) = (1/2,1/2)^T$.

Observe that for all $x \in [x^1,x^2]$, $\nabla f(x)$ lies in the cone generated by $n_1$ and $n_2$. Adding constraint 2 to the active set yields

$$A = \{1,2\}\ ,\ q = 2, \qquad H = \begin{bmatrix} 0 & 0 & 0 \\ 0 & 0 & 0 \\ 0 & 0 & 1 \end{bmatrix}, \text{ and } \quad N^* = \begin{bmatrix} 1/2 & -1 & 0 \\ 3/2 & -2 & 0 \end{bmatrix}$$

Iteration 3: $x^2$ is the optimal solution of $P(\{1,2\})$. Compute $s(x^2) = (0,0,-8)^T$. The only possible choice for p is 3. Now compute $z=(0,0,1)^T$, $r=(2,4)^T$, $t_1 = \min\left\{\frac{1/2}{2}, \frac{1/2}{4}\right\} = 1/8$, $z^Tn_3 = 1$, and $t_2=8$.

Since $t_1 < t_2$ set $t = t_1$ and take a partial step; i.e., compute

$x^3 = x^2 + tz = (-1,4,1/8)^T$

$f = 1 + 1/8 \cdot 1\ (1/16 + 0) = 129/128$

and $u = (1/2,\ 1/2,\ 0)^T + 1/8(-2,-4,1)^T = (1/4,0,1/8)^T$.

Observe that $\nabla f(x)$ lies in the cone generated by $n_1$, $n_2$ and $n_3$ for all points $x \in [x^2,x^3]$. It lies outside of this cone at all points on the semi-infinite ray $\{x|=x^2+tz,\ t>t_1\}$ and, in particular, at the point $x^2+t_2z$ that would have been reached by a full step.

Dropping constraint 2 from the active set yields $A = \{1\}$, $q = 1$,

$$H = \frac{1}{25}\begin{bmatrix} 9 & -12 & 0 \\ -12 & 16 & 0 \\ 0 & 0 & 25 \end{bmatrix} \quad \text{and } N^* = \frac{1}{25}(-4,-3,0).$$

Iteration 4: Compute $z = Hn_3 = 1/25(24,-32,25)$, $r=N^*n_3 = 6/25$, $t_1 = 25/24$, $z^Tn^3 = 89/25$, $s_3(x^3) = -63/8$ and $t_2 = 1575/712$.

Since $t_1 < t_2$, set $t = t_1$ and take a partial step; i.e., compute

$x^4 = x^3 + tz = (0,8/3,7/6)^T$

$f = 129/128 + 25/24 \cdot 89/25\ (25/48 + 1/8) = 245/72$

and $u = (1/4,\ 1/8)^T + 25/24(-6/25,1)^T = (0,7/6)^T$.

A remark analogous to the one made in the previous iteration concerning $\nabla f(x)$ can be made here as well.

Dropping constraint 1 from the active set yields $A = \emptyset$, $q=0$, and $H=I$.

Iteration 5: Since $A = \emptyset$ we take a full step and compute

$$z=Hn_3 = (0,-2,1),\ t = t_2 = -s(x^4)/z^Tn_3 = \frac{25/6}{5} = 5/6.$$

$$x^5 = x^4 + tz = (0,1,2,)^T$$

$$f = 245/72 + 5/6 \cdot 5(5/12+7/6) = 10$$

and $\quad u = 7/6 + 5/6 = 2.$

Adding constraint 3 to the active set yields

$$A = \{3\}\ ,\ q = 1,\quad H = 1/5\begin{bmatrix} 5 & 0 & 0 \\ 0 & 1 & 2 \\ 0 & 2 & 4 \end{bmatrix} \text{ and } N^* - 1/5(0,-2,1).$$

Iteration 6: $x^5$ is the optimal solution of $P(\{3\})$.

Compute $s(x^5) = (5,-1,0)^T$. The only possible choice for p is 2. Now compute

$z = Hn_2 = (2,1/5,2/5)^T$, $z^T n_2 = 21/5$, $r = -2/5$ $t_1 = \infty$, and $t_2 = 5/21$.

Since $t_2 < t_1$, we set $t = t_2$, take a full step and compute

$$x^6 = x^5 + tz = 1/21(10,22,44)^T$$

$$f = 10 + 5/21 \cdot 21/5(5/42 + 0) = 425/42$$

and $u = (2,0)^T + 5/21(2/5,1)^T = 1/21(44,5)^T$

Adding constraint 2 to the active set yields

$$A = \{3,2\}, \; q = 2, \; H = 1/21 \begin{bmatrix} 1 & -2 & -4 \\ -2 & 4 & 8 \\ -4 & 8 & 16 \end{bmatrix}, \text{ and } N^* = 1/21 \begin{bmatrix} 4 & -8 & 5 \\ 10 & 1 & 2 \end{bmatrix}$$

Iteration 7: $x^6$ is the optimal solution of $P(\{3,2\})$.

Since $s(x^6) = (62/21,0,0)^T \geqslant 0$, $x^6$ is also the optimal solution for the full problem.

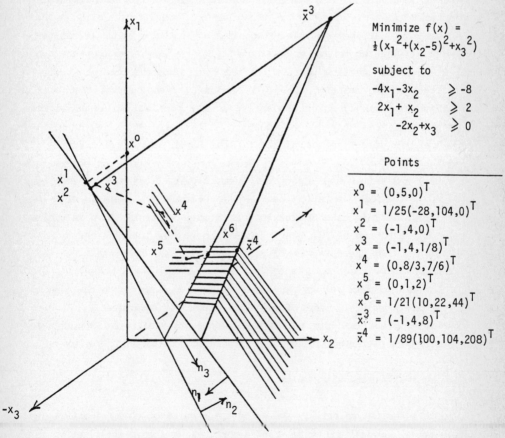

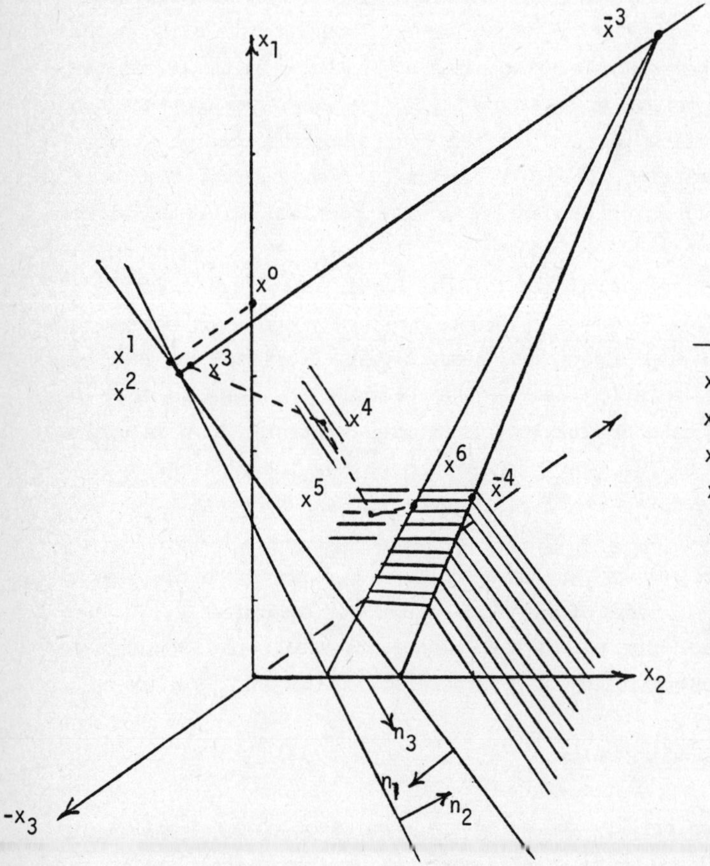


Minimize $f(x) = \frac{1}{2}(x_1^2 + (x_2-5)^2 + x_3^2)$

subject to

$$-4x_1 - 3x_2 \geqslant -8$$
$$2x_1 + x_2 \geqslant 2$$
$$-2x_2 + x_3 \geqslant 0$$

| Points |
|---|
| $x^0 = (0,5,0)^T$ |
| $x^1 = 1/25(-28,104,0)^T$ |
| $x^2 = (-1,4,0)^T$ |
| $x^3 = (-1,4,1/8)^T$ |
| $x^4 = (0,8/3,7/6)^T$ |
| $x^5 = (0,1,2)^T$ |
| $x^6 = 1/21(10,22,44)^T$ |
| $\bar{x}^3 = (-1,4,8)^T$ |
| $\bar{x}^4 = 1/89(100,104,208)^T$ |

Figure 1: An example showing different solution paths for the dual and primal-dual algorithms.

Primal-Dual Solution

The primal-dual algorithm performs identically to the dual algorithm until iteration 3. Since it always takes a full dual step, the primal-dual algorithm computes

$t = t_2 = 8$

$x^3 = x^2 + tz = (-1,4,8)^T$

$f = 1 + 8 \cdot 1(4 + 0) = 33$

and $u = (1/2,1/2,0)^T + 8(-2,-4,1)^T = (-15/2, -31/2\ 8)^T$ at iteration 3. Since two components of u are negative $x^3$ is not the optimal solution of $P(\{1,2,3\})$. When solving $P(\{1,2,3\})$ the primal part of the primal-dual algorithm has a choice of which constraint to drop from the active set at $x^3$. If it drops constraint 1 it obtains $x^6$, the optimal solution to the full problem in one step. If it drops constraint 2, it first moves to $x^4 = 1/89(100,104,208)$, the minimum of f(x) subject to constraint 1 and 3 treated as equalities. It then drops constraint 1 and adds back constraint 2 to obtain $x^6$.

Observe that whichever of these two paths are followed by the primal part of the primal-dual algorithm it obtains $P(\{3,2\})$ and $x^6$ as the next subproblem - optimal solution pair, while the dual algorithm obtains a different subproblem and solution, $P(\{3\})$ and $x^5$. Also observe, that at $x^5$ the second constraint, which was active at the optimal solution of the previous subproblem, has become violated. This can never happen in the primal-dual case since the solution of $P(A \cup \{p\})$ must satisfy all constraints indexed by A. Let $P(A_D)$ and $P(A_{PD})$ denote the new subproblems determined by the dual and primal-dual algorithms, respectively, starting from the same subproblem - optimal solution pair. Our example above shows that $A_D \not\supseteq A_{PD}$ is possible. This can only happen when some constraint in A is violated by the solution of $P(A_D)$ as happened above. Now let us suppose that $A_{PD} \subsetneqq A_D$ and that the solutions of $P(A_D)$ and $P(A_{PD})$ are different. But this is impossible since then $A_{PD} \subsetneqq A_D \subseteq (A \cup \{p\})$ and $P(A_{PD}) = P(A_D) = P(A \cup \{p\})$.

In actual practice, it is very uncommon for $P(A_D)$ not to be the same as $P(A_{PD})$. Consequently, in terms of major iterations (or equivalently, subproblem - optimal solution pairs) both algorithms usually follow the same path, their only differences being of a rather local character.

## 4. Summary of Computational Results

In this section we briefly summarize the results of computational tests that we ran on our dual and primal-dual algorithms, comparing their performance against that of primal algorithms. A more detailed presentation and analysis of these results is given in [6] and [9]. Two different kinds of computational experiments were performed.

In the first set of experiments we randomly generated 24 different types of strictly convex quadratic programming problems with known optimal solutions using the technique of Rosen and Suzuki [13]. Each problem type was determined by specifying the number of variables m (9,27,or 81), the number of constraints k (m,or 3m), the number of constraints q* in the active set A at the solution (k/9 or k/3), and the condition (well or ill) of the Hessian matrix G. To generate G its off-diagonal elements were set to r(-1,1) and $G_{11}=S_1+r(0,1)+1$ was computed, where r(a,b) denotes a freshly computed (pseudo-) random number uniformly distributed between a and b and $S_i$ denotes the sum of the absolute values of the off-diagonal elements in the i-th row of G. In the well-conditioned case we set $G_{ii}=S_i+r(0,1)+1$, for i=2,...,m, while in the ill-conditioned case we set $G_{ii} = G_{i-1}+S_i+S_{i-1}+r(0,1)$, for i=2,...,m.

Further, our experiments were subdivided into three runs. In runs 1 and 2 all 16 problem types with m equal to 9 and 27 were generated and the optimal dual variables $u_j$, $j \in A$ were set to r(0,30) and r(0,30k), respectively. In run 3, all 8 problem types with m equal to 81 were generated, and we set $u_j$=r(0,81k) for $j \in A$. To complete the generation of each problem we set the components of the optimal solution x* to r(-5,5) and the elements of C to r(-1,1). The columns of C were then normalized to unit length. For $j \in A$ we set $s_j=0$ and for $j \notin A$ we set $s_j = r(0,1)$ and $u_j = 0$. Then we set $b=s-C^Tx^*$ and a = Cu-Gx*.

The test results reported in Tables 1 and 2 were for versions of the dual and primal-dual algorithms in which the most violated constraint was always chosen in step (1). The primal algorithms that we used for comparisons were algorithms 1 and 2 given by Goldfarb [5]. When applied to a strictly convex quadratic programming problem, algorithm 1 is identical to the algorithm given by Fletcher [4]. For finding a feasible point for the primal algorithms we used the variant of Rosen's [12] procedure suggested in [5], where the operators $N^+ = (N^TN)^{-1}N^T$ and $P = I-NN^+$ used by Rosen are replaced by N* and H, defined by (2) and (3). Powell [11] recently showed that Rosen's feasible point method can cycle. No cycling, however, was encountered in any of our runs. The feasible point routine was started from a randomly generated point. All of the algorithms used the same matrix factorizations and updating procedures mentioned near the end of section 2, and were coded modularly in FORTRAN so that they all used many of the same subroutines.

In Table 1 the number of operations relative to the dual was obtained by computing the ratio of the number of operations (multiplications + divisions + 10*square roots) required by a given method compared with that required by the dual method for all replications of a particular problem type and then averaging this ratio over all problem types for that run. Each problem type was replicated five times in runs 1 and 2 and once in run 3.

| | | | Number of Operations Relative To The Dual | | | |
|---|---|---|---|---|---|---|
| RUN | Number of Variables | Number of Problems | Feasible Point | Primal - Alg. 1 | Phase 2 Alg. 2 | Primal - Dual |
| 1 | 9,27 | 80 | 1.6 | 3.2 | 1.7 | 1.2 |
| 2 | 9,27 | 80 | 1.5 | 3.0 | 1.6 | 1.3 |
| 3 | 81 | 8 | 1.5 | 3.8 | 1.7 | -.- - - |
| Average | | | 1.5 | 3.3 | 1.7 | 1.3 |
| Phase 1 + Phase 2 | | | | 4.8 | 3.2 | |

Table 1: Comparison of Algorithms: Number of Operations Relative to the Dual.

On the average our algorithm for finding a feasible point required 50 percent more operations (see Table 1) and 49 percent more basis change (see Table 2) than did the dual algorithm. In fact the dual algorithm required fewer operations than the feasible point algorithm on all problem types except those with (m,k,q*, condition of G) equal to (9,27,9, well and ill) in runs 1 and 2, (27,81,27,well) and (9,9,3,ill) in run 2, and (81,243,81,well) in run 3. In no case did either of the combined feasible point-primal algorithms require less than one and one-half as many operations as did the dual algorithm. As one would expect, the problem types that were the most difficult for the dual to solve were those where the optimal solution occurred at a vertex, where the optimal dual variables were large as in runs 2 and 3, and where G was well-conditioned. This set of conditions ensured that the unconstrained optimum was very distant from the constrained optimum. Even in these cases the ratios of the total number of operations required by the feasible point routine and algorithm 2 (the better of the two primal algorithms) to that required by the dual were 1.51 for m=9, 1.49 for m=27 and 1.92 for m=81. On the average the best feasible point-primal combination required approximately three times as much computation (see Table 1) and basis changes (see Table 2) as the dual algorithm.

One reason for the superior performance of the dual algorithm is that it does not appear to add many constraints to the active set that are not in the final basis. The figures in parentheses in the third and fourth columns in Table 2 give respectively the number of problems in which the dual algorithm never dropped a constraint from the active set and the theoretical mininum number of basis changes required in total by the dual algorithm on all problems in that run (which is achieved if no constraint is added to the active set that later has to be dropped). The difference between the number of basis changes for the dual and this theoretical minimum equals tow times the number of "wrong" constraints added by the method at intermediate steps. As Table 2 shows in addition to 1568 "correct" constraints which were added during the course of our experiments, 422 "wrong" constraints were also added to the basis (and later dropped from it). Of the latter more thant 60 percent of the "wrong" additions occurred in the solution of the three problem types mentioned above as being the most difficult for the dual method.

| Run | Number of Variables | Number of Problems | Dual | Number of Basis Changes Feasible Point | Primal - Alg.1 | Phase 2 Alg.2 | Primal -Dual |
|---|---|---|---|---|---|---|---|
| 1 | 9,27 | 80,(62) | 689(640) | 1498 | 1716 | 1506 | 693 |
| 2 | 9,27 | 80(46) | 1128(640) | 1498 | 1882 | 1538 | 1168 |
| 3 | 81 | 8(2) | 594(288) | 674 | 1284 | 996 | -- |
| TOTALS: | | 168(110) | 2411(1568) | 3670 | 4882 | 4040 | — |
| Ratio to Dual: | | | | 1.49 | 1.98 | 1.64 | 1.02 |

Table 2: Comparison of Algorithms: The Number of Basis Changes

Test results are also given for the primal-dual algorithm in Table 1 and 2. As the primal part of this algorithm we used algorithm 1. In all problems except those in which the optimal solution occurred at a vertex this algorithm behaved identical to the dual except that on major iterations which required dropping constraints from this basis the primal-dual algorithm added the violated constraint p before, rather than after, dropping the appropriate constraints. On the problems with a vertex solution in primal-dual algorithm always required as many as or more steps and basis changes than the dual algorithm because it would drop and then add back the same constraint to the active set during a call to the primal routine. The situation illustrated by the example in section 3 occurred only once in the 160 problem runs. Consequently it appears that except in very rare cases the dual and primal-dual algorithms given in section 2 will proceed through the same sequence of subproblem-optimal solution pairs.

In the second set of experiments we used our dual algorithm to solve the sequences of QP problems generated by Powell´s successive quadratic programming algorithm [10] (implemented as Harwell Subroutine VFO2AD) in the course of its solution of six nonlinear programming test problems. The performance of our algorithm was compared with the performance of Fletcher´s feasible point-primal QP codes [2,3] that are normally used by Powell´s VFO2AD code. This feasible point routine determines a feasible vertex and is more closely related to a standard phase 1 simplex algorithm than it is to the feasible point routine described earlier in this section. Fletcher´s primal QP algorithm is identical to primal algorithm 1 when applied to strictly convex QP problems except for its implementation; e.g., the matrices H and N* rather than factorizations are stored and updated. The results of our computational experiments are summarized in Table 3.

| Problem | Number of Variables | Number of Constraints | Aver. q* | Number of Basis Changes Feas.Pt. | Primal | Dual | Ratio of Operations to Dual Feas.Pt. | Primal | Total |
|---|---|---|---|---|---|---|---|---|---|
| Powell | 5(6) | 3*(5) | 4.0 | 4 | 12 | 24 | .38 | 1.47 | 1.85 |
| POP | 3(4) | 7 (9) | 2.2 | 30 | 11 | 13 | 1.19 | 1.87 | 3.06 |
| Triangle | 7(8) | 9 (11) | 6.1 | 66 | 43 | 83 | .75 | 2.46 | 3.21 |
| Colville 1 | 5(6) | 15(17) | 6.0 | 28 | 10 | 34 | .81 | 1.66 | 2.47 |
| Colville 2 | 15(16) | 20(22) | 14.0 | 118 | 52 | 239 | .76 | 2.57 | 3.33 |
| Colville 3 | 5(6) | 16(18) | 6.0 | 16 | 0 | 28 | .75 | 1.44 | 2.19 |
| Average | | | | | | | .77 | 1.91 | 2.69 |

Table 3. Comparison of Algorithms when used as Subroutines in Powell's Successive Quadratic Programming Code.

The numbers of variables and constraints for the generated Quadratic Programming problems (given in parentheses in Table 3) are respectively one and two more than these numbers for the original nonlinear problems. To obtain the results in this table the number of basis changes and operations were summed over all QP problems generated during the solution a given nonlinear problem. Observe that although now on the average the feasible point routine is less expensive than the dual, the total work for it and Fletcher's primal algorithm is 2.69 times that required by the dual algorithm. This is in spite of the fact that the optimal solutions of all of the QP problems solved were on manifolds of dimension less than or equal to two. In Colville 1 and 3 all optimal solutions occurred at vertices; (see the columns with headings "Variables" and "Average q*.) A complete description of the use of the dual algorithm as a subroutine in Powell's algorithm and a more complete description of the numerical results is given in [6,9].

## References.


1. Biggs, M. C. (1975) "Constrained minimization using recursive quadratic programming: some alternative subproblem formulations" in Towards global optimization, eds. L.C.W. Dixon and G.P. Szego, North-Holland Publishing Co. (Amsterdam).

2. Fletcher, R. (1970). "The Calculation of Feasible Points for Linearly Constrained Optimization Problems", UKAEA Research Group Report, AERE R 6354 (Harwell).

3. Fletcher, R. (1970). "A FORTRAN Subroutine for Quadratic Programming". UKAEA Research Group Report. AERE R 6370 (Harwell).

4. Fletcher, R. (1971)."A general quadratic programming algorithm" Journal Inst. Math. Applics.,Vol. 7, pp. 76-91.

5. Goldfarb, D. (1972). "Extension of Newton's method and simplex methods for solving quadratic program", in Numerical Methods for Nonlinear Optimization, ed. F. Lootsma, Academic Press (London), pp.239-254

6. Goldfarb, D. and Idnani, A. U.(1981) "A numerically stable dual method for solving strictly convex quadratic programs". The City College of New York, Department of Computer Sciences. Technical Report 81-102, (New York)

7. Han, S-P (1976) "Superlinearly convergent variable metric algorithms for general nonlinear programming problems", Mathematical Programming, Vol. 11, pp.263-282.

8. Han, S-P (1977) "A globally convergent method for nonlinear programming", Journal of Optimization Theory and Applications, Vol.22, pp. 297-309.

9. Idnani, A.U (1980). "Numerically stable dual projection methods for solving positive definite quadratic programs." Ph.D. Thesis, The City College of New York, Department of Computer Sciences (New York).

10. Powell, M.J.D. (1978) "A fast algorithm for nonlinearly constrained optimization calculations" in Numerical Analysis, Dundee, 1977 Lecture Notes in Mathematics 630 (Springer Verlag, Berlin) pp.144-157.

11. Powell, M.J.D., (1980) "An example of cycling in a feasible point algorithm", Report 1980/NA5 DAMTP, University of Cambridge, (Cambridge, England).

12. Rosen, J.B. (1960) "The gradient projection method for nonlinear programming, Part 1. Linear constraints", SIAM Journal of Applied Math. Vol. 8, pp. 181-217.

13. Rosen, J. B. and Suzuki, S. (1965) "Construction of nonlinear programming test problems", Communications of the ACM, pp. 113.

The design and use of a frontal scheme for solving sparse unsymmetric equations

Iain S. Duff
Computer Science and Systems Division
AERE Harwell, Didcot, Oxon OX11 ORA.

Abstract

We first describe frontal schemes for the solution of large sparse sets of linear equations and then discuss the implementation of a code in the Harwell Subroutine Library which solves unsymmetric systems using this approach. We indicate the performance of our software on some test examples.

## 1. Introduction

This paper discusses the design of Harwell subroutine MA32 for the solution of sets of linear equations whose coefficient matrix is large, sparse and unsymmetric. The method used is the frontal method. We discuss the history of this method and our code in this introduction and describe the algorithm for frontal schemes in section 2. In section 3, we discuss our implementation particularly emphasising novel features in our code. Finally, we comment on the performance of our code in section 4.

Irons (1970) is generally accredited with the first publication of a code for implementing a frontal solution method. His code was only designed for symmetric positive definite systems and did no pivoting. A frontal code for unsymmetric systems was published by Hood (1976) but contained many deficiencies and inefficiencies. Most of these were remedied by Cliffe et al (1978) upon whose programs our work is based.

It is our belief that ours is the first unsymmetric frontal code which meets the standards required for incorporation in a general purpose mathematical software library. Our code has been placed in the Harwell Subroutine Library under the generic name MA32 (Duff (1981)).

## 2. Frontal schemes

The basis for all frontal schemes is Gaussian elimination. That is, we perform the LU decomposition of a permutation of A which we can write as

$$A = PL.UQ \tag{1}$$

where P,Q are permutation matrices and L and U are lower and upper triangular matrices respectively. Because of the size of the problems we are considering, PL and UQ will ordinarily be held on an auxiliary storage device.

An important observation is that only the factors PL and UQ are used during the solution process and, in frontal schemes, we make use of this by never storing (or indeed generating) the whole of A at one time.

Although frontal schemes were originally developed for the solution of finite element discretizations in structural analysis (Irons (1970)) where the resulting assembled stiffness matrix is positive definite, they are really applicable to a far wider class of problems and can be modified to work when the resulting matrix is indefinite or even unsymmetric. Indeed the software we will describe in this paper can be used to solve any general unsymmetric set of linear equations although it will not always be the most efficient method.

However, it is easiest to describe the frontal method by reference to its application in the solution of a finite element problem. The crucial observation is illustrated in Figure 1 where the operation on entry (i,j) of the matrix at one step

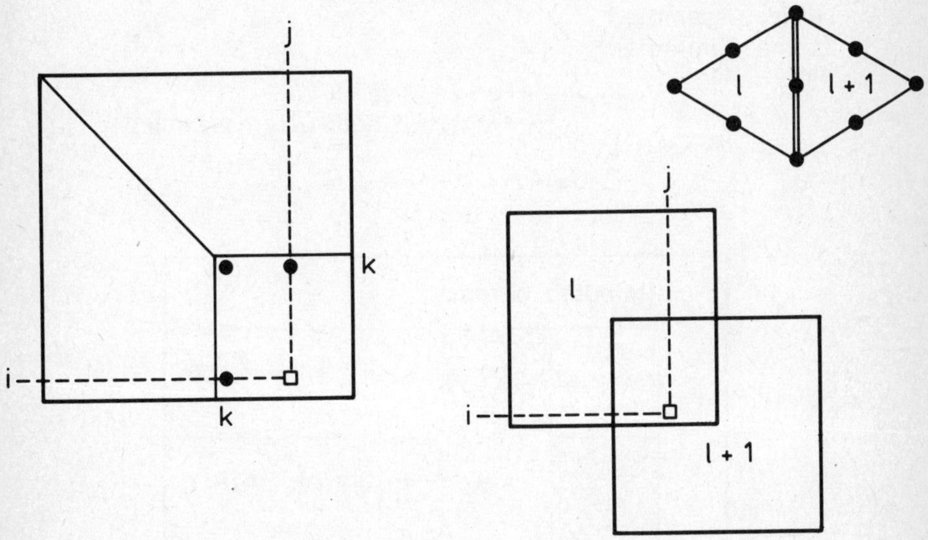


1(a) Elimination. 1(b) Assembly.

Figure 1 Elimination and assembly in finite element problems

of Gaussian elimination is of the form

$$a_{ij} \leftarrow a_{ij} - a_{ik}[a_{kk}]^{-1} a_{kj} \tag{2}$$

while the basic assembly operation in a finite element calculation (see Figure 1(b)) is of the form

$$a_{ij} \leftarrow a_{ij} + e_{65}^{(\ell)} + e_{32}^{(\ell+1)} \quad . \tag{3}$$

Now, so long as the entries in the triple product in (2) are all fully summed (i.e. there are no more contributions of the form (3) to these entries), it does not matter in what order the computations (2) and (3) are performed.

We utilise this observation by making a judicious choice for ordering the assembly of the contributions from the finite elements and by concurrently assembling and eliminating so that the size of our partially assembled submatrix is kept low. We illustrate this in Figure 2, where the triangular elements are numbered according to the order in which they are assembled. The matrix in Figure 2(b)

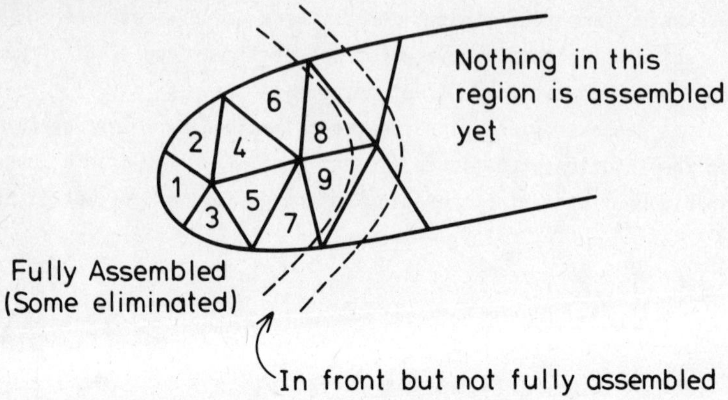


2(a) Underlying Problem.

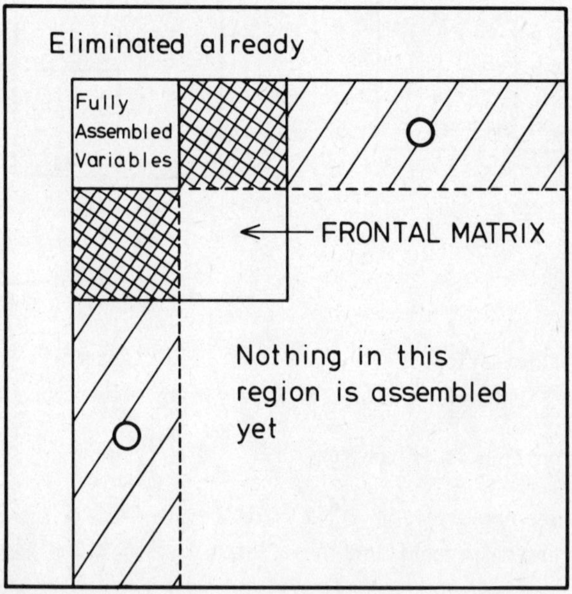


2(b) Matrix after assembly of element 9.

Figure 2 Midway through a frontal scheme.

represents the situation after the assembly of element 9. The zero blocks in Figure 2(b) illustrate clearly that we can perform eliminations, choosing pivots from anywhere within the block of fully-assembled variables, containing our computations entirely within the frontal matrix.

In the absence of any need for numerical pivoting (for example, when the assembled matrix is positive definite) the submatrix corresponding to the fully-assembled variables could be eliminated and, since these rows and columns take no further part in the factorization of the matrix, they could be written to in-core buffers or sent to backing store. Since the lower right part of the matrix in Figure 2(b) is not yet even partially assembled, the only part that need be held in core at this time is the frontal matrix.

The size of the front will of course vary as the assembly and elimination progresses. For any geometry and ordering on the elements, there is clearly a maximum front size and our space allocation for the two-dimensional frontal matrix must be sufficient to hold this maximum.

This economisation of storage is the main reason for the popularity and use of frontal methods since a judicious ordering of the elements can often reduce the maximum front-width to a small fraction of the total size of the problem particularly if the discretized region is long and thin. Although we cannot expect to reduce the number of operations or storage for the factors to that of a general sparse code (for example, Duff (1977)), all operations of the form (2) will be performed using ordinary full matrix code. The avoidance of indirect addressing at this innermost loop means that the time per operation will be much lower than for a general sparse code and that vectorization will be facilitated on machines capable of it.

The frontal technique can also be applied to systems which are not positive definite. If we return to Figure 2(b), it is evident that pivots can be chosen from anywhere within the fully-assembled block. We can thus add the requirement that the pivot satisfies some numerical test. In this present code, we use the same criterion as in our general sparse codes, namely threshold pivoting. That is, entry $a_{\ell k}$ will be regarded as suitable for use as a pivot only if

$$|a_{\ell k}| \geqslant u.\max_i |a_{ik}| \qquad (4)$$

where u is some user set number between 0 and 1 and the maximum is over all rows in the fully-assembled block and the hatched region of Figure 2(b). Although we may reach a stage when large entries in the hatched region prevent any entry from the fully-assembled block being numerically suitable for use as a pivot, this is no problem. We simply perform some more assemblies to cause those large entries in the hatched region to move into the fully-assembled block where they can then be pivotal. The only penalty we pay is that our frontal matrix must accommodate these extra rows and columns and so will be larger than if no pivoting were required.

Although this description was based on problems arising from finite element discretizations, it is just as easy to use such techniques where the matrix is fully assembled and is input by rows. The situation is illustrated in Figure 3(b) where we show the frontal matrix after the input of equation 3 from a five-point discretization of the Laplacian operator on a 2x4 grid, as shown in Figure 3(a). At this stage no further equations will cause any non-zero entries to appear in the

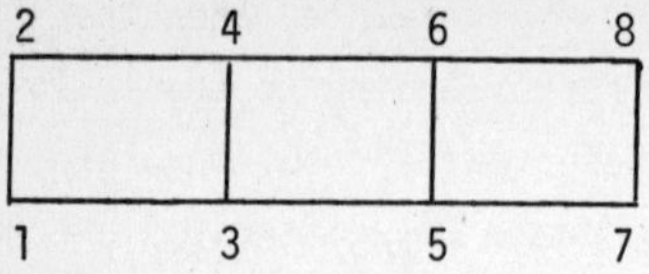


3(a) 2x4 grid

$$\begin{array}{cccc} -4 & 1 & 1 & \\ 1 & -4 & 0 & 1 \\ 1 & 0 & -4 & 1 \end{array}$$

3(b) First three rows of matrix

Figure 3 Illustration of equation input in frontal schemes

first column of Figure 3(b), so this column is effectively fully summed and any entry in it (subject perhaps to a numerical criterion) can be chosen as pivot. In this application, the frontal matrix will be rectangular rather than square and the large number of zero entries in it indicates an overhead which might be considerable if the method were applied to quite general systems.

## 3. Implementation

There are two minor difficulties which any frontal code must resolve. The first is that when we perform the backsubstitution operations on our modified right hand sides we wish to use the matrix factors in the reverse direction to that in which they were generated. To facilitate this, we use direct access data sets to hold the factors since, on many systems, and particularly on our IBM, backspacing is very expensive. In order to reduce the I/O overheads to these data sets (one is used for PL and the other for the UQ of (1)), the factors are blocked, and it is these blocks which are written from or to an in-core buffer. Since different manufacturers implement direct access in different ways, this part of the code is necessarily system dependent but we have isolated these dependencies in a few subroutines.

The second minor difficulty is that the algorithm is not clairvoyant and so is unable to tell when a variable is fully assembled. We therefore perform a prepass on the structure only and generate a single vector which records, for each variable, the element (or equation) in which it appears for the last time.

We now list some of the novel features of our code and expand on one or two of these in the rest of this section.

(i) The subroutine structure has been designed for greater modularity and isolates access to auxiliary storage.

(ii) System dependent routines are used to perform dynamic file allocation and formatting in the IBM version of the code. These are clearly identified to facilitate conversion for other machines.

(iii) An option exists for the user to input his data by equations rather than by elements.

(iv) With the exception of the I/O routines, all subroutines are in portable Fortran (1966 ANSI standard) and pass the PFORT verifier. Numerous data checks have been incorporated to make the software robust. The user level subroutines subdivide workspace so that the user need declare only a single workspace array of each type.

(v) New internal data structures have been used to facilitate the assembly operations.

(vi) A new pivoting strategy is used and a "static condensation" facility has been incorporated.

(vii) Useful information is returned in the event of failure due to insufficient space for the frontal matrix or the matrix factors.

(viii) Reverse communication has been used to give greater flexibility to the user interface.

These are all discussed in detail by Duff (1981). We will discuss (vi), (vii) and (viii) in the following.

Our basic pivoting strategy is to use threshold pivoting as indicated in (4). The threshold value (u) is held as a common block variable set to a default value of 0.1 and so can be reset by the user. It is possible that our front size is increased because none of our fully-summed variables satisfy the threshold criterion. Should this increase cause the front size to be larger than that allocated to the in-core frontal matrix, then entries closest to satisfying the numerical test will be chosen in order that the elimination may proceed. In this case, a flag will be set to warn the user of possible instability. In some instances, we may wish to sacrifice a little stability for the sake of efficient pivot selection. The user may stipulate the maximum number of fully-summed variables which can stay in the frontal matrix and sufficient eliminations will always be performed to ensure that this number is not exceeded even if some of the pivots chosen do not satisfy the tolerance. In conjunction with this, the number of fully-summed columns searched for a pivot can also be restricted. We had originally removed this option from the code of Cliffe et al (1978) but were persuaded to reinstate it by Jackson who quoted instances of problems whose solution was only made feasible by the use of such controls. Many finite element formulations include variables which are internal to the element and can thus be eliminated without reference to any other elements. It is much more efficient to perform these "static condensations" within the element itself rather than after assembling the element into the frontal matrix. We have therefore incorporated this option into our code and have found cases (Cliffe, private communication) where savings of over 30% in execution time have been obtained.

The user must provide values for the order of the frontal matrix and must allocate storage for the matrix factors. If the problem is complicated or unfamiliar it may be hard to choose appropriate values, so we feel it is very important to return useful information should the run fail due to insufficient space. This is easy in the case of storage for the factors. We simply continue the decomposition throwing away the factors (that is, overwriting information in the buffers) and calculating the space required for subsequent runs. If insufficient space is allocated to the frontal matrix, the situation is slightly more complicated.

In this case, we continue with a symbolic factorization only whose space requirements are linear rather than quadratic in the front size. We can thus continue so long as the front size does not exceed the square of the order of the frontal matrix and can return the order of frontal matrix required. Although this will be sufficient for subsequent runs on the same data, the user may wish to increase it slightly to allow for numerical pivoting. In both cases, a flag is set so that the run can be optionally terminated immediately the allocated space is exceeded.

Such flexibility in user control is facilitated by the use of reverse communication. This means that control is returned to the user each time an assembly operation is required. As we indicated earlier, this assembly can be the input of a finite element (and, optionally, the right hand sides) or could be an equation (row) of the coefficient matrix of any set of unsymmetric equations. Thus, the call structure is of the form shown in Figure 4.

```
for each element or equation do
  begin
    input element or equation;
    call frontal solver;
  end
```

Figure 4 Use of reverse communication with frontal solver

This structure permits the generation of elements or equations in a way most convenient to the user and also allows regular interaction with the frontal solver, as was illustrated at the end of the preceding paragraph.

## 4. Performance

We illustrate the performance of our frontal code, MA32, on a model problem by comparing it with a general solver for unsymmetric matrices (MA28, Duff (1977)) on 5-point discretizations of the Laplacian operator on rectangular grids. We show these results in Table 1. We see that core requirements of the frontal code are much less than for the general code and its execution time is quite competitive with MA28-FACTOR and much better than the MA28-ANALYZE. However, the real power of the frontal method is most evident on large finite element problems and we illustrate its performance on such problems in Table 2.

These runs were performed by Cliffe et al (1978) and arose in the study of flow in a cavity. The total time (inclusive of I/O operations) can be seen to be about 1.5 μsec per Gaussian elimination operation which is less than 20% above the basic machine time on the 370/168 for such operations.

| Grid m<br>n | 10<br>10 | 10<br>40 | 10<br>60 | 10<br>100 | 10<br>300 | 32<br>32 | 64<br>64 |
|---|---|---|---|---|---|---|---|
| Decomposition time | | | | | | | |
| MA32A | 60 | 240 | 370 | 610 | 1800 | 2000 | 23000 |
| MA28* | 100 | 560 | 850 | 1410 | 5000 | 8700 | N.A. |
| MA28** | 30 | 130 | 210 | 350 | 1100 | 1100 | N.A. |
| Solution time | | | | | | | |
| MA32 | 10 | 45 | 60 | 100 | 310 | 250 | 1900 |
| MA28 | 5 | 15 | 20 | 40 | 120 | 60 | 300+ |
| Storage in kbytes | | | | | | | |
| MA32 | 20 | 20 | 20 | 20 | 20 | 50 | 150 |
| MA28 | 15 | 70 | 110 | 180 | 530 | 320 | 1300+ |

Table 1 Times (in msecs on an IBM 3033) on a model problem on an mxn grid.

+Estimated

*Time for pivot selection and factorization (MA28-ANALYZE)

**Time for factorization after pivot sequence is known (MA28-FACTOR)

| Number of elements | Number of nodes | Number of degrees of freedom (Order) | Frontwidth | Time |
|---|---|---|---|---|
| 1596 | 3317 | 7495 | 145 | 225 |
| 228 | 509 | 1159 | 55 | 6 |
| 69 | 315 | 709 | 35 | 2 |
| 165 | 663 | 2817 | 77 | 22 |
| 1568 | 3249 | 3249 | 60 | 22 |

Table 2 Runs on large finite element problems (Cliffe et al (1978)). The frontwidth is the maximum but is attained for much of the computation. The times are in seconds on an IBM 370/168.

## References

Cliffe, K.A., Jackson, C.P., Rae, J. and Winters, K.H. (1978). Finite element flow modelling using velocity and pressure variables. Harwell Report, AERE R.9202.

Duff, I.S. (1977). MA28 - a set of Fortran subroutines for sparse unsymmetric linear equations. Harwell Report, AERE R.8730, HMSO, London.

Duff, I.S. (1981). MA32-A package for solving sparse unsymmetric systems using the frontal method. Harwell Report, AERE R.10079, HMSO, London.

Hood, P. (1976). Frontal solution program for unsymmetric matrices. Int. J. Numer. Meth. Engng. 10, pp.379-399.

Irons, B.M. (1970). A frontal solution program for finite element analysis. Int. J. Numer. Meth. Engng. 2, pp.5-32.

Vol. 759: R. L. Epstein, Degrees of Unsolvability: Structure and Theory. XIV, 216 pages. 1979.

Vol. 760: H.-O. Georgii, Canonical Gibbs Measures. VIII, 190 pages. 1979.

Vol. 761: K. Johannson, Homotopy Equivalences of 3-Manifolds with Boundaries. 2, 303 pages. 1979.

Vol. 762: D. H. Sattinger, Group Theoretic Methods in Bifurcation Theory. V, 241 pages. 1979.

Vol. 763: Algebraic Topology, Aarhus 1978. Proceedings, 1978. Edited by J. L. Dupont and H. Madsen. VI, 695 pages. 1979.

Vol. 764: B. Srinivasan, Representations of Finite Chevalley Groups. XI, 177 pages. 1979.

Vol. 765: Padé Approximation and its Applications. Proceedings, 1979. Edited by L. Wuytack. VI, 392 pages. 1979.

Vol. 766: T. tom Dieck, Transformation Groups and Representation Theory. VIII, 309 pages. 1979.

Vol. 767: M. Namba, Families of Meromorphic Functions on Compact Riemann Surfaces. XII, 284 pages. 1979.

Vol. 768: R. S. Doran and J. Wichmann, Approximate Identities and Factorization in Banach Modules. X, 305 pages. 1979.

Vol. 769: J. Flum, M. Ziegler, Topological Model Theory. X, 151 pages. 1980.

Vol. 770: Séminaire Bourbaki vol. 1978/79 Exposés 525–542. IV, 341 pages. 1980.

Vol. 771: Approximation Methods for Navier-Stokes Problems. Proceedings, 1979. Edited by R. Rautmann. XVI, 581 pages. 1980.

Vol. 772: J. P. Levine, Algebraic Structure of Knot Modules. XI, 104 pages. 1980.

Vol. 773: Numerical Analysis. Proceedings, 1979. Edited by G. A. Watson. X, 184 pages. 1980.

Vol. 774: R. Azencott, Y. Guivarc'h, R. F. Gundy, Ecole d'Eté de Probabilités de Saint-Flour VIII-1978. Edited by P. L. Hennequin. XIII, 334 pages. 1980.

Vol. 775: Geometric Methods in Mathematical Physics. Proceedings, 1979. Edited by G. Kaiser and J. E. Marsden. VII, 257 pages. 1980.

Vol. 776: B. Gross, Arithmetic on Elliptic Curves with Complex Multiplication. V, 95 pages. 1980.

Vol. 777: Séminaire sur les Singularités des Surfaces. Proceedings, 1976-1977. Edited by M. Demazure, H. Pinkham and B. Teissier. IX, 339 pages. 1980.

Vol. 778: SK1 von Schiefkörpern. Proceedings, 1976. Edited by P. Draxl and M. Kneser. II, 124 pages. 1980.

Vol. 779: Euclidean Harmonic Analysis. Proceedings, 1979. Edited by J. J. Benedetto. III, 177 pages. 1980.

Vol. 780: L. Schwartz, Semi-Martingales sur des Variétés, et Martingales Conformes sur des Variétés Analytiques Complexes. XV, 132 pages. 1980.

Vol. 781: Harmonic Analysis Iraklion 1978. Proceedings 1978. Edited by N. Petridis, S. K. Pichorides and N. Varopoulos. V, 213 pages. 1980.

Vol. 782: Bifurcation and Nonlinear Eigenvalue Problems. Proceedings, 1978. Edited by C. Bardos, J. M. Lasry and M. Schatzman. VIII, 296 pages. 1980.

Vol. 783: A. Dinghas, Wertverteilung meromorpher Funktionen in ein- und mehrfach zusammenhängenden Gebieten. Edited by R. Nevanlinna and C. Andreian Cazacu. XIII, 145 pages. 1980.

Vol. 784: Séminaire de Probabilités XIV. Proceedings, 1978/79. Edited by J. Azéma and M. Yor. VIII, 546 pages. 1980.

Vol. 785: W. M. Schmidt, Diophantine Approximation. X, 299 pages. 1980.

Vol. 786: I. J. Maddox, Infinite Matrices of Operators. V, 122 pages. 1980.

Vol. 787: Potential Theory, Copenhagen 1979. Proceedings, 1979. Edited by C. Berg, G. Forst and B. Fuglede. VIII, 319 pages. 1980.

Vol. 788: Topology Symposium, Siegen 1979. Proceedings, 1979. Edited by U. Koschorke and W. D. Neumann. VIII, 495 pages. 1980.

Vol. 789: J. E. Humphreys, Arithmetic Groups. VII, 158 pages. 1980.

Vol. 790: W. Dicks, Groups, Trees and Projective Modules. IX, 127 pages. 1980.

Vol. 791: K. W. Bauer and S. Ruscheweyh, Differential Operators for Partial Differential Equations and Function Theoretic Applications. V, 258 pages. 1980.

Vol. 792: Geometry and Differential Geometry. Proceedings, 1979. Edited by R. Artzy and I. Vaisman. VI, 443 pages. 1980.

Vol. 793: J. Renault, A Groupoid Approach to C*-Algebras. III, 160 pages. 1980.

Vol. 794: Measure Theory, Oberwolfach 1979. Proceedings 1979. Edited by D. Kölzow. XV, 573 pages. 1980.

Vol. 795: Séminaire d'Algèbre Paul Dubreil et Marie-Paule Malliavin. Proceedings 1979. Edited by M. P. Malliavin. V, 433 pages. 1980.

Vol. 796: C. Constantinescu, Duality in Measure Theory. IV, 197 pages. 1980.

Vol. 797: S. Mäki, The Determination of Units in Real Cyclic Sextic Fields. III, 198 pages. 1980.

Vol. 798: Analytic Functions, Kozubnik 1979. Proceedings. Edited by J. Ławrynowicz. X, 476 pages. 1980.

Vol. 799: Functional Differential Equations and Bifurcation. Proceedings 1979. Edited by A. F. Izé. XXII, 409 pages. 1980.

Vol. 800: M.-F. Vignéras, Arithmétique des Algèbres de Quaternions. VII, 169 pages. 1980.

Vol. 801: K. Floret, Weakly Compact Sets. VII, 123 pages. 1980.

Vol. 802: J. Bair, R. Fourneau, Etude Géometrique des Espaces Vectoriels II. VII, 283 pages. 1980.

Vol. 803: F.-Y. Maeda, Dirichlet Integrals on Harmonic Spaces. X, 180 pages. 1980.

Vol. 804: M. Matsuda, First Order Algebraic Differential Equations. VII, 111 pages. 1980.

Vol. 805: O. Kowalski, Generalized Symmetric Spaces. XII, 187 pages. 1980.

Vol. 806: Burnside Groups. Proceedings, 1977. Edited by J. L. Mennicke. V, 274 pages. 1980.

Vol. 807: Fonctions de Plusieurs Variables Complexes IV. Proceedings, 1979. Edited by F. Norguet. IX, 198 pages. 1980.

Vol. 808: G. Maury et J. Raynaud, Ordres Maximaux au Sens de K. Asano. VIII, 192 pages. 1980.

Vol. 809: I. Gumowski and Ch. Mira, Recurences and Discrete Dynamic Systems. VI, 272 pages. 1980.

Vol. 810: Geometrical Approaches to Differential Equations. Proceedings 1979. Edited by R. Martini. VII, 339 pages. 1980.

Vol. 811: D. Normann, Recursion on the Countable Functionals. VIII, 191 pages. 1980.

Vol. 812: Y. Namikawa, Toroidal Compactification of Siegel Spaces. VIII, 162 pages. 1980.

Vol. 813: A. Campillo, Algebroid Curves in Positive Characteristic. V, 168 pages. 1980.

Vol. 814: Séminaire de Théorie du Potentiel, Paris, No. 5. Proceedings. Edited by F. Hirsch et G. Mokobodzki. IV, 239 pages. 1980.

Vol. 815: P. J. Slodowy, Simple Singularities and Simple Algebraic Groups. XI, 175 pages. 1980.

Vol. 816: L. Stoica, Local Operators and Markov Processes. VIII, 104 pages. 1980.

Vol. 817: L. Gerritzen, M. van der Put, Schottky Groups and Mumford Curves. VIII, 317 pages. 1980.

Vol. 818: S. Montgomery, Fixed Rings of Finite Automorphism Groups of Associative Rings. VII, 126 pages. 1980.

Vol. 819: Global Theory of Dynamical Systems. Proceedings, 1979. Edited by Z. Nitecki and C. Robinson. IX, 499 pages. 1980.

Vol. 820: W. Abikoff, The Real Analytic Theory of Teichmüller Space. VII, 144 pages. 1980.

Vol. 821: Statistique non Paramétrique Asymptotique. Proceedings, 1979. Edited by J.-P. Raoult. VII, 175 pages. 1980.

Vol. 822: Séminaire Pierre Lelong-Henri Skoda, (Analyse) Années 1978/79. Proceedings. Edited by P. Lelong et H. Skoda. VIII, 356 pages, 1980.

Vol. 823: J. Král, Integral Operators in Potential Theory. III, 171 pages. 1980.

Vol. 824: D. Frank Hsu, Cyclic Neofields and Combinatorial Designs. VI, 230 pages. 1980.

Vol. 825: Ring Theory, Antwerp 1980. Proceedings. Edited by F. van Oystaeyen. VII, 209 pages. 1980.

Vol. 826: Ph. G. Ciarlet et P. Rabier, Les Equations de von Kármán. VI, 181 pages. 1980.

Vol. 827: Ordinary and Partial Differential Equations. Proceedings, 1978. Edited by W. N. Everitt. XVI, 271 pages. 1980.

Vol. 828: Probability Theory on Vector Spaces II. Proceedings, 1979. Edited by A. Weron. XIII, 324 pages. 1980.

Vol. 829: Combinatorial Mathematics VII. Proceedings, 1979. Edited by R. W. Robinson et al.. X, 256 pages. 1980.

Vol. 830: J. A. Green, Polynomial Representations of $GL_n$. VI, 118 pages. 1980.

Vol. 831: Representation Theory I. Proceedings, 1979. Edited by V. Dlab and P. Gabriel. XIV, 373 pages. 1980.

Vol. 832: Representation Theory II. Proceedings, 1979. Edited by V. Dlab and P. Gabriel. XIV, 673 pages. 1980.

Vol. 833: Th. Jeulin, Semi-Martingales et Grossissement d'une Filtration. IX, 142 Seiten. 1980.

Vol. 834: Model Theory of Algebra and Arithmetic. Proceedings, 1979. Edited by L. Pacholski, J. Wierzejewski, and A. J. Wilkie. VI, 410 pages. 1980.

Vol. 835: H. Zieschang, E. Vogt and H.-D. Coldewey, Surfaces and Planar Discontinuous Groups. X, 334 pages. 1980.

Vol. 836: Differential Geometrical Methods in Mathematical Physics. Proceedings, 1979. Edited by P. L. García, A. Pérez-Rendón, and J. M. Souriau. XII, 538 pages. 1980.

Vol. 837: J. Meixner, F. W. Schäfke and G. Wolf, Mathieu Functions and Spheroidal Functions and their Mathematical Foundations Further Studies. VII, 126 pages. 1980.

Vol. 838: Global Differential Geometry and Global Analysis. Proceedings 1979. Edited by D. Ferus et al. XI, 299 pages. 1981.

Vol. 839: Cabal Seminar 77 – 79. Proceedings. Edited by A. S. Kechris, D. A. Martin and Y. N. Moschovakis. V, 274 pages. 1981.

Vol. 840: D. Henry, Geometric Theory of Semilinear Parabolic Equations. IV, 348 pages. 1981.

Vol. 841: A. Haraux, Nonlinear Evolution Equations- Global Behaviour of Solutions. XII, 313 pages. 1981.

Vol. 842: Séminaire Bourbaki vol. 1979/80. Exposés 543-560. IV, 317 pages. 1981.

Vol. 843: Functional Analysis, Holomorphy, and Approximation Theory. Proceedings. Edited by S. Machado. VI, 636 pages. 1981.

Vol. 844: Groupe de Brauer. Proceedings. Edited by M. Kervaire and M. Ojanguren. VII, 274 pages. 1981.

Vol. 845: A. Tannenbaum, Invariance and System Theory: Algeb and Geometric Aspects. X, 161 pages. 1981.

Vol. 846: Ordinary and Partial Differential Equations, Proceedi Edited by W. N. Everitt and B. D. Sleeman. XIV, 384 pages. 1

Vol. 847: U. Koschorke, Vector Fields and Other Vector Bu Morphisms – A Singularity Approach. IV, 304 pages. 1981.

Vol. 848: Algebra, Carbondale 1980. Proceedings. Ed. by R Amayo. VI, 298 pages. 1981.

Vol. 849: P. Major, Multiple Wiener-Itô Integrals. VII, 127 pages. 1

Vol. 850: Séminaire de Probabilités XV. 1979/80. Avec table géné des exposés de 1966/67 à 1978/79. Edited by J. Azéma and M. IV, 704 pages. 1981.

Vol. 851: Stochastic Integrals. Proceedings, 1980. Edited by Williams. IX, 540 pages. 1981.

Vol. 852: L. Schwartz, Geometry and Probability in Banach Spac X, 101 pages. 1981.

Vol. 853: N. Boboc, G. Bucur, A. Cornea, Order and Convexit Potential Theory: H-Cones. IV, 286 pages. 1981.

Vol. 854: Algebraic K-Theory. Evanston 1980. Proceedings. Edi by E. M. Friedlander and M. R. Stein. V, 517 pages. 1981.

Vol. 855: Semigroups. Proceedings 1978. Edited by H. Jürgens M. Petrich and H. J. Weinert. V, 221 pages. 1981.

Vol. 856: R. Lascar, Propagation des Singularités des Soluti d'Equations Pseudo-Différentielles à Caractéristiques de Multi cités Variables. VIII, 237 pages. 1981.

Vol. 857: M. Miyanishi. Non-complete Algebraic Surfaces. XV 244 pages. 1981.

Vol. 858: E. A. Coddington, H. S. V. de Snoo: Regular Boundary Va Problems Associated with Pairs of Ordinary Differential Expressio V, 225 pages. 1981.

Vol. 859: Logic Year 1979–80. Proceedings. Edited by M. Lerm J. Schmerl and R. Soare. VIII, 326 pages. 1981.

Vol. 860: Probability in Banach Spaces III. Proceedings, 1980. Edi by A. Beck. VI, 329 pages. 1981.

Vol. 861: Analytical Methods in Probability Theory. Proceedings 19 Edited by D. Dugué, E. Lukacs, V. K. Rohatgi. X, 183 pages. 19

Vol. 862: Algebraic Geometry. Proceedings 1980. Edited by A. L gober and P. Wagreich. V, 281 pages. 1981.

Vol. 863: Processus Aléatoires à Deux Indices. Proceedings, 19 Edited by H. Korezlioglu, G. Mazziotto and J. Szpirglas. V, 274 pag 1981.

Vol. 864: Complex Analysis and Spectral Theory. Proceedin 1979/80. Edited by V. P. Havin and N. K. Nikol'skii, VI, 480 pag 1981.

Vol. 865: R. W. Bruggeman, Fourier Coefficients of Automorp Forms. III, 201 pages. 1981.

Vol. 866: J.-M. Bismut, Mécanique Aléatoire. XVI, 563 pages. 19

Vol. 867: Séminaire d'Algèbre Paul Dubreil et Marie-Paule Malliav Proceedings, 1980. Edited by M.-P. Malliavin. V, 476 pages. 19

Vol. 868: Surfaces Algébriques. Proceedings 1976–78. Edited J. Giraud, L. Illusie et M. Raynaud. V, 314 pages. 1981.

Vol. 869: A. V. Zelevinsky, Representations of Finite Classical Grou IV, 184 pages. 1981.

Vol. 870: Shape Theory and Geometric Topology. Proceedings, 19 Edited by S. Mardešić and J. Segal. V, 265 pages. 1981.

Vol. 871: Continuous Lattices. Proceedings, 1979. Edited by B. Ban schewski and R.-E. Hoffmann. X, 413 pages. 1981.

Vol. 872: Set Theory and Model Theory. Proceedings, 1979. Edite by R. B. Jensen and A. Prestel. V, 174 pages. 1981.